STATISTICAL DATA ANALYSIS FOR THE PHYSICAL SCIENCES

Data analysis lies at the heart of every experimental science. Providing a modern introduction to statistics, this book is ideal for undergraduates in physics. It introduces the necessary tools required to analyse data from experiments across a range of areas, making it a valuable resource for students.

In addition to covering the basic topics, the book also takes in advanced and modern subjects, such as neural networks, decision trees, fitting techniques and issues concerning limit or interval setting. Worked examples and case studies illustrate the techniques presented, and end-of-chapter exercises help test the reader's understanding of the material.

ADRIAN BEVAN is a Reader in Particle Physics in the School of Physics and Astronomy, Queen Mary, University of London. He is an expert in quark flavour physics and has been analysing experimental data for over 15 years.

STATISTICAL DATA ANALYSIS FOR THE PHYSICAL SCIENCES

ADRIAN BEVAN

Queen Mary, University of London

CAMBRIDGE
UNIVERSITY PRESS

CAMBRIDGE UNIVERSITY PRESS

Cambridge, New York, Melbourne, Madrid, Cape Town,
Singapore, São Paulo, Delhi, Mexico City

Cambridge University Press
The Edinburgh Building, Cambridge CB2 8RU, UK

Published in the United States of America by Cambridge University Press, New York

www.cambridge.org
Information on this title: www.cambridge.org/9781107670341

First published 2013

Printed and bound in the United Kingdom by the MPG Books Group

A catalogue record for this publication is available from the British Library

ISBN 978-1-107-03001-5 Hardback
ISBN 978-1-107-67034-1 Paperback

Contents

Preface

The foundations of science are built upon centuries of careful observation. These constitute measurements that are interpreted in terms of hypotheses, models, and ultimately well-tested theories that may stand the test of time for only a few years or for centuries. In order to understand what a single measurement means we need to appreciate a diverse range of statistical methods. Without such an appreciation it would be impossible for scientific method to turn observations of nature into theories that describe the behaviour of the Universe from sub-atomic to cosmic scales. In other words science would be impracticable without statistical data analysis. The data analysis principles underpinning scientific method pervade our everyday lives, from the use of statistics we are subjected to through advertising to the smooth operation of SPAM filters that we take for granted as we read our e-mail. These methods also impact upon the wider economy, as some areas of the financial industry use data mining and other statistical techniques to predict trading performance or to perform risk analysis for insurance purposes.

This book evolved from a one-semester advanced undergraduate course on statistical data analysis for physics students at Queen Mary, University of London with the aim of covering the rudimentary techniques required for many disciplines, as well as some of the more advanced topics that can be employed when dealing with limited data samples. This has been written by a physicist with a non-specialist audience in mind. This is not a statistics book for statisticians, and references have been provided for the interested reader to refer to for more rigorous treatment of the techniques discussed here. As a result this book provides an up-to-date introduction to a wide range of methods and concepts that are needed in order to analyse data. Thus this book is a mixture of a traditional text book approach and a teach by example approach. By providing these opposing viewpoints it is hoped that the reader will find the material more accessible. Throughout the book, a number of case studies are presented with possible solutions discussed in detail. The purpose of these sections is to consolidate the more abstract notions discussed in the book and apply them to an example. In some instances the case study may appear somewhat abstract and specific to scientific research; however, where possible more widely applicable problems have been included. At the end of each chapter there is a summary of the main issues raised, followed by a number of example questions to help the reader practise and gain a deeper understanding of the material included. Solutions to questions are presented at the end of the book.

The Introduction motivates the importance of studying statistical methods when analysing data by looking at three common problems encountered early within the life of a physicist: measuring g, testing Ohm's law and studying the law of radioactive decay. Following this motivational introduction the book is divided into two parts: (i) the foundations of statistical data analysis from set notation through to confidence intervals, and (ii) discussion of more advanced topics in the form of optimisation, fitting, and data mining. The material in the first part of the book is ordered logically so that successive sections build on material discussed in the earlier ones, while the second part of the book contains stand alone chapters that depend on concepts developed in the first part. These later chapters can be read in any order.

The first part of this book starts with an introduction to sets and Venn diagrams that provide some of the language that we use to discuss data. Having developed this language, the concept of probability is formally introduced in Chapter 3. Readers who are familiar with these concepts already may wish to skip over the first two chapters and proceed straight to the discussion in Chapter 4 on how to visualise and quantify data. Distributions of data are often described by simple functions that are used to represent the probability of observing data with a certain value. A number of useful distributions are described in Chapter 5, and Appendix B builds on this topic by discussing a number of additional functions that may be of use. Measurements are based on the determination of some central value of an observable quantity, with an uncertainty or error on that observable. Issues surrounding uncertainties and errors are introduced in Chapter 6, and this topic is further developed in Chapter 7. Chapter 8 discusses hypothesis testing and brings together many of the earlier concepts in the book.

The second part of the book presents more advanced topics. Chapter 9 discusses fitting data given some assumed model using χ^2 and likelihood methods. This relies heavily on concepts developed in Chapters 5 and 6, and Appendix B. Chapter 10 discusses data mining, or how to efficiently separate two classes of data, for example signal from background using numerical methods. The methods discussed include the use of 'cut-based' selection, the Bayesian classifier, Fisher's linear discriminant, artificial neural networks, and decision trees.

To avoid interrupting the flow of the text, a number of detailed appendices have been prepared. The most important of these appendices is a collection of probability tables, which is conveniently located at the end of the book in order to provide a quick reference to the reader. There is also a glossary of terms intended to help the reader when referring back to the book some time after an initial reading. Appendices listing a number of commonly used probability density functions, and elementary numerical integration techniques have also been provided. While these are not strictly required in order to understand the concepts introduced in the book, they have been included in order to make this a more complete resource for readers who wish to study this topic beyond an undergraduate course.

There are a number of technical terms introduced throughout this book. When a new term is introduced, that term is highlighted in ***bold-italic text*** to help the reader refer back to this description at a later time.

I would like to thank colleagues who have provided me with feedback on the draft of this book, and in particular Peter Crew.

for the pendulum is given by

$$T = 2\pi \sqrt{\frac{L}{g}}. \tag{1.1}$$

This is valid for small oscillations, as the small angle approximation of $\sin\theta \simeq \theta$ is used in deriving the relationship between T and g. By using Eq. (1.1) it is possible to estimate the acceleration due to the Earth's gravity from measurements of (i) the length L and (ii) the period of oscillation T of the pendulum. There is no dependence on the amplitude of oscillation (as long as the small-angle approximation remains valid) or the mass of the bob at the end of the pendulum. Equation (1.1) can be re-arranged as follows

$$g = \frac{4\pi^2 L}{T^2}, \tag{1.2}$$

so that one can directly obtain g from a single measurement or data point. Propagation of errors is discussed in Chapter 6 where in particular Eq. (6.11) can be used in order to determine the uncertainty on g, denoted by σ_g, given measurements and uncertainties on both T and L, where

$$\sigma_g^2 = \left(\frac{8\pi^2 L}{T^3}\right)^2 \sigma_T^2 + \left(\frac{4\pi^2}{T^2}\right)^2 \sigma_L^2. \tag{1.3}$$

Although this is a common experiment performed in many schools and undergraduate laboratories, many aspects of data analysis are required to fully appreciate issues that may arise with the measurement of g.

The most straightforward way to approach this problem is to measure the time it takes for a single complete oscillation to occur. From some maximum displacement one can release the pendulum bob and measure the time taken for the bob to reach back to where it started from. This measurement neglects any small effect arising from air resistance that may reduce the amplitude of oscillation slightly. There are several factors that one should consider when performing a measurement of g using this method.

- If a stopwatch is used to measure the period of oscillation, then there will be a significant uncertainty associated with starting and stopping the watch, relative to the period of time. For example if the oscillation period is of the order of a second, then the reaction time of the person starting and stopping the watch in quick succession will play an important role in the accuracy and precision of the time period measurement. One needs to ensure firstly that the measurement is accurate (i.e. that there is no systematic mistake made in timing), and that it is sufficiently precise. If the method of measuring the time period has an uncertainty of a second, then it would not be possible to make a useful measurement of time periods less than one second.

For example one can use the standard deviation on the ability of the experimenter to stop a stopwatch at a given count as a measure of the uncertainty of timing a particular event. As one has to both start and stop the stopwatch, twice this uncertainty can be ascribed as the uncertainty on time measurement. Using a trial of ten attempts to count to ten seconds on a stopwatch, the RMS deviation from that number was found to be 0.1 s. This was taken as the uncertainty on starting or stopping the watch. As both of these events have the same source of uncertainty they are taken to be correlated, hence twice that is used as the uncertainty on timing for an individual measurement of the oscillation i.e. ± 0.2 s.

- The relative uncertainty on L should also be small. For example, a measurement made with $L = 5.0 \pm 0.5$ cm would introduce a 10% uncertainty in Eq. (1.3), whereas this can be reduced to the percent level by increasing L to 50 cm. That in turn will increase T hence the relative precision on T using a given timing device as $T \propto \sqrt{L}$. So the experiment should be designed in such a way that L is sufficiently large so that it is not a dominant factor in the total uncertainty obtained for g, and mitigates contributions from timing.
- The relationship given in Eq. (1.1) is valid only for small amplitudes of oscillation, hence any measurement that deviates from a small amplitude of oscillation will result in a biased determination of g. The experimenter needs to understand this assumption and how the underlying approximation restricts the maximum displacement of the pendulum from the vertical position. For example, a length of $L = 50$ cm with an amplitude of oscillation of 10 cm results in a bias of -0.7% on the angle θ, which in turn introduces a small bias on g. As the effect is non-linear, an amplitude of oscillation of 20 cm results in a bias of -2.7% on the angle used in the approximation and so on. Therefore when starting to swing the pendulum, one should take care that the amplitude of oscillation is sufficiently small that the small angle approximation remains valid. A longer string length will help minimise this source of bias.

Having determined that the uncertainty on measurement timing is 0.2 s using a stopwatch, then a period of oscillation lasting 2 s will be determined to 10% (0.2 s /2.0 s). This in turn will limit the precision with which g can be determined as can be seen from the first term in Eq. (1.3). Table 1.1 shows several measurements of g made by using a stopwatch for timing and $L = 64.6 \pm 0.5$ cm. The measurement obtained using a single oscillation is $g = (9.7 \pm 2.4)$ m/s^2. The relative precision of this measurement is quite poor, only having determined g to 25%; however, it is possible to reduce the uncertainty on the measurement in several different

Table 1.1 *Measurements of g using the methodology described in the text.*

Number of measurements	Number of oscillations	g (m/s^2)
1	1	9.7 ± 2.4
10	1	10.2 ± 0.5
1	10	9.7 ± 0.3
10	10	9.7 ± 0.1

ways using the same experimental apparatus, but varying the experimental method slightly.

- One can make several measurements, and take the average of the ensemble as an estimate of the mean value of the period of oscillation, with the standard deviation corresponding to a measure of the spread in the data as the uncertainty on a measurement. The uncertainty on the mean value of the period from N individual measurements will scale by a factor $1/\sqrt{N}$. This means that one can quickly make improvements on a single measurement, by making several subsequent measurements, but soon the increase in precision obtained by making an additional single measurement will become small. This assumes that each trial measurement is made under identical conditions, and neglects any systematic mistakes in measuring L or T.
- Another way to improve the precision would be to measure the time taken for several periods of oscillation. The uncertainty on a single period of oscillation derived from a measurement of ten oscillations is $\sigma_T = \sigma_{10T}/10$, i.e. the uncertainty on the measurement is spread equally across each oscillation that occurred within the measurement, and one can effectively reduce the statistical uncertainty on T by a factor of ten.
- The period of oscillation achievable for a given setup depends on the length L. Given that g is a constant for a given laboratory, the longer the pendulum, the longer the period of oscillation. So one can increase L for a given measurement method in order to reduce the relative precision on the measured value of g, within practical limitations of the experimental apparatus.

The precision obtained for the measured value of g can be improved further by averaging a number of measurements made of multiple oscillations at a time, for the longest pendulum length L allowable by the the apparatus, i.e. by taking into account the three previous considerations. Table 1.1 summarises the results of several sets of measurements of g and in particular illustrates the potential for improvements obtained by making multiple measurements, and by measuring several periods of oscillation. One can see from the results shown in the table that while making several measurements of a given number of oscillations and averaging those results gives an improvement in precision over a single measurement of g, the most effective way to improve the precision using this apparatus is to increase the number of oscillations counted in order to estimate T. By applying a basic understanding of statistical data analysis to this problem it has been possible to refine the experimental method in order to determine the value of g with a relative precision of 1%, compared with the 25% relative precision obtained initially. For a measurement with a 1% statistical precision, a systematic bias of -0.7% from the use of the small angle approximation would be a concern. The underlying techniques used here are discussed in Chapters 4 and 6 and the reader may wish to re-read this example once they have reached the end of Chapter 6.

1.2 Verification of Ohm's law

This example builds on some of the techniques discussed above. We are surrounded by electronic devices in the modern world. One of the fundamental laws related to electronics is that of Ohm's law: the voltage V across an Ohmic conductor is proportional to the electric current I passing through it. The constant of proportionality is called the resistance of the conductor, and components called resistors that are made out of Ohmic conductors pervade our lives in numerous ways. The electronic circuits in your mobile phone, television, and computer have hundreds of resistors in them, and without the simple resistor those devices would cease to function. The underlying principle of Ohm's law underpinning the concept of the resistor is

$$V = IR. \tag{1.4}$$

Given the form of this relationship it is possible to take a single measurement of V and I and subsequently compute an estimate of R. This gives the resistance of the conductor for a given data point, but does not allow the experimenter to verify if the conductor is Ohmic.

The measurement of a single data point depends on the precision with which the voltage and current were measured. As an example we can consider measurements made on a $12\,000\,\Omega$ resistor using a hand-held digital multi-meter (DMM) to determine the voltage across the resistor, and a precision DMM to determine the current passing through the resistor. In this case the limiting factor is the voltage measurement, which was made with a precision of 0.09% and an additional

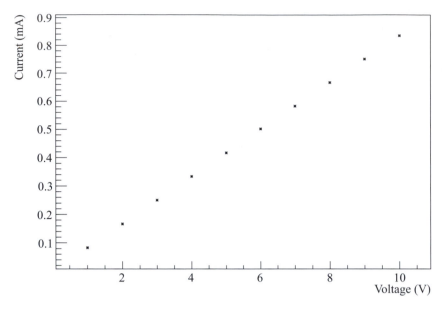

Figure 1.1 The distribution of current (I) passing through a $12\,000\,\Omega$ resistor versus the voltage (V) across it.

uncertainty of two in the last digit read off the DMM.[1] The current was measured with a more precise device, so it is sufficient to assume that the uncertainty on a single measurement of R discussed here is dominated by the contribution from the voltage measurement. The resistance measured, with a current of 0.8349 mA passing through it and a voltage of 10.0 V across it, is $11\,977 \pm 35\,\Omega$.

If one assumes that the conductor was known to be Ohmic, which is reasonable for a resistor, then one could simply average the values of resistance obtained for a number of different measurements. As with the example of measuring g above, the mean and standard deviation of the data could be used to determine an estimate of the resistance and uncertainty of the component under study (see Chapter 4). The mean resistance computed for a $12\,000\,\Omega$ resistor as obtained from the ten data points shown in Figure 1.1 is $11\,996\,\Omega$, with a standard deviation of $30\,\Omega$. This is in good agreement with the estimate of the central value from a single measurement. However, in general the precision from ten measurements should be $\sqrt{10}$ times better (smaller) than that of a single measurement. As this is not the case with these data one might worry that there could be systematic effect (such as linearity of a measuring device, or temperature of the laboratory) that is not being taken into

---

[1] DMMs need to be regularly calibrated in order to ensure that measurements made are accurate, and that the precisions of measurements as quoted in their instruction manuals are valid. There may also be systematic effects that one has to consider, such as temperature dependences on a reported reading. The accuracy and limitations of a measuring device used needs to be understood in order to ensure that the experimenter can correctly interpret the data recorded.

account. More generally if it is not known if the component is an Ohmic conductor one could plot I against V and determine if Eq. (1.4) was valid or not. Figure 1.1 shows data recorded for a $12\,000\,\Omega$ resistor over a voltage range of $1{-}10\,V$. The linear relationship between I and V is evident by eye.

The value of R measured for this resistor could be determined more precisely by fitting the data as described in Chapter 9. A χ^2 fit to V/I vs V assuming Ohm's law yields $R = 11\,996 \pm 10\,\Omega$, in good agreement with previous determinations. The χ^2 for this fit is 8.87 for nine degrees of freedom, which corresponds to a probability of $P(\chi^2, \nu) = P(8.87, 9) = 0.45$ which is quite reasonable (see Chapter 5). If one looks at the data in Figure 1.1, one might wonder if the value of R might be changing slightly with voltage. This can be investigated by taking data over a wide range of V, or using the existing data and changing the model used in the fit. The next simplest model to Ohm's law would be to introduce a linear variation in the resistance as a function of voltage. A χ^2 fit to V/I vs V assuming a relationship of the form of $y = mx + R$ to allow for a possible change in resistance as a function of voltage yields $R = 12\,037 \pm 20\,\Omega$ and $m = -7.5 \pm 3.3$ with $\chi^2 = 3.75$ for eight degrees of freedom. The slope m obtained is consistent with zero within uncertainties, and the value of R obtained is consistent with the previous determination, but has a larger uncertainty. The probability of this fit is quite reasonable, $P(\chi^2, \nu) = P(3.75, 8) = 0.878$. While the result of the second fit is more probable than the first, there is insufficient motivation to support the hypothesis that the behaviour observed deviates from Ohm's law, as the slope coefficient obtained from the second fit is consistent with zero. Examining the data using techniques such as this allows one to go beyond a qualitative inspection of graphs and to quantify the underlying physical observables. Chapter 9 discusses several different fitting techniques that could have been applied to this problem.

Looking closely at the data it appears that the value of R computed for the first data point is slightly higher than the rest, and as a result will be the main contributor to the value of the χ^2 obtained when testing Ohm's law. While this data point is reasonable one might be concerned about the integrity of the data in general. In such a situation there are several options that one might naturally consider: (i) taking more data points at lower values of V, (ii) repeating the full set of measurements, and (iii) studying the specifications of the measuring devices to ensure that the uncertainties have correctly interpreted when making each of the measurements.

1.3 Measuring the half-life of an isotope

A common undergraduate experiment involves the determination of the decay constant (λ) or half-life ($t_{1/2} = \ln 2/\lambda$) of a radioactive isotope. The number of

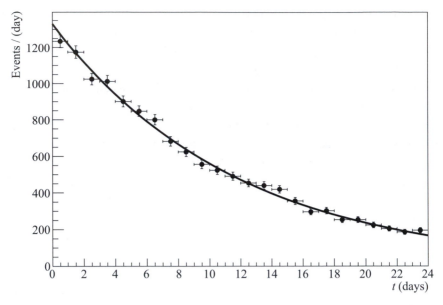

Figure 1.2 The distribution of events (count rate) expected in an experiment studying the radioactive decay of ^{131}I.

radioactive nuclei at a given time t is

$$N(t) = N_0 e^{-\lambda t}, \tag{1.5}$$

where N_0 is the initial number of nuclei in the sample (at time $t = 0$). The rate of decay of a radioactive isotope is given by

$$\frac{dN}{dt} = -\lambda N(t). \tag{1.6}$$

The decay constant λ introduced above can be understood as the rate of change of the number of radioactive nuclei of a given type with respect to time elapsed. This quantity is related to the aptly called half-life of an isotope. The *half-life* is the time taken for the number of radioactive nuclei to reduce by one half with respect to a given time. It follows from Eq. (1.5) that after one half-life

$$\frac{N(t)}{N_0} = e^{-\lambda t_{1/2}} = \frac{1}{2}, \tag{1.7}$$

hence $t_{1/2} = \ln 2/\lambda$.

It is possible to measure the decay constant of a radioactive isotope by studying the number of counts observed in a radiation detector, for example a Geiger–Müller tube, as a function of time. Each count corresponds to the detection of the decay product of a radioactive nuclei disintegrating. From the expected time dependence one can extract the decay constant and in turn convert this into a measure of the half-life of the isotope. Figure 1.2 shows the result of a simulation of the count rate expected as a function of time for an ^{131}I source, which produces β radiation with a half-life of 8.02 days. This isotope is used in a number of medical physics applications. The simulation neglects background, which can be measured in the absence of the source and subtracted from data, or alternatively taken into account at the same time as extracting λ from the data. Background from the naturally occurring radioactivity of the experimental environment is expected to be uniform in time. There are a number of possible sources of background including cosmic rays and natural radioactivity from rocks or other materials in the environment, e.g. areas rich in granite often have elevated levels of radon gas which is radioactive, and thus a marginally higher than normal level of background radiation. While this background varies from location to location, for a given laboratory the background rate should be constant.

The ^{131}I signal shown in the figure follows the radioactive decay law of Eq. (1.5), with an exponentially decreasing count rate. The data are displayed as a binned histogram (see Chapter 4), with Poisson error bars[2] on the content of each of the bins (see Chapters 6 and 7). As the number of entries in a given bin (so the count rate) decrease, so the relative size of the error on the count rate increases. The data are fitted using an un-binned extended maximum likelihood fit as described in Chapter 9 using a model corresponding to a signal exponentially decaying with time (i.e. the ^{131}I). So while the data are visually displayed in bins of counts in any given day, the individual time of a count is used in the fit to data, and the binning is essentially for display purposes only. The results of this fit to data will be the values and uncertainties of the signal yield (so how many signal counts were recorded), and the decay constant measured for the ^{131}I sample. Given the relationship between the decay constant and half-life, it is possible to convert the value of λ obtained in the fit to data into a measurement of the half-life using the error propagation formalism introduced in Chapter 6. In order to avoid having to translate the fitted parameter into the half-life, one can re-parameterise the likelihood function to fit for $t_{1/2}$ directly.

This is just one way to analyse the data, instead of performing an un-binned extended maximum likelihood fit to the data, we could have binned the data before fitting. On doing this some information is lost, but if there are sufficient data to analyse, any loss in precision would be negligible. Another alternative would be to perform a χ^2 fit to data (as was used for the example of studying Ohmic conductors discussed above). Yet another way to analyse the data would be to perform a linear least squares regression analysis. In order to do this it is convenient to convert the data into a form where one has a linear relationship between the quantities plotted on the ordinate and abscissa. Given that the measured count rate is

---

[2] The process of detecting the products of a radioactive decay is described by a Poisson probability distribution (see Chapter 5 for more details), and so the uncertainty ascribed to the content of a given bin is Poisson.

given by

$$N(t) = N_0 e^{-\lambda t} + N_{bg} U(t), \tag{1.8}$$

where $U(t)$ is a uniform probability density function (see Appendix B) describing some number of background events N_{bg}, it follows that the background corrected rate $N'(t) = N(t) - N_{bg}$ satisfies

$$\ln N'(t) = \ln N_0 - \lambda t. \tag{1.9}$$

Hence, one can determine the decay constant from the slope of the logarithm $\ln N'(t)$ vs time, and one can compute the value and uncertainty on λ without the need for sophisticated minimisation techniques that underpin χ^2 or maximum-likelihood fitting approaches. In addition, the initial total number of radioactive nuclei can be computed from the constant term, $\ln N_0$, should this be of interest. All of the techniques mentioned here are described in Chapter 9. The reader may wish to re-examine this example once they have reached the end of Chapter 9 in order to reflect on the approach taken, and on how some of the alternate techniques mentioned above may be applied to this problem.

A number of physical processes obey an exponential decay (or growth) law where the data can be analysed in a similar way to that described here. For example measurement of the lifetime of the decay of a sub-atomic particle, such as muons found in cosmic rays, uses the same data analysis technique(s) as the decay constant or half-life analysis discussed here. Similarly data obtained in order to determine the attenuation of light in a transparent material, or radioactive particles passing through different thicknesses of shielding follow an exponential attenuation law and can be analysed using one of the approaches described here. The basis for this type of data analysis can also be adapted in order to address more complicated situations where the physical model involves more parameters (λ, N_0 and N_{bg} are physical parameters associated with this particular problem), more than one signal or background component, or indeed more dimensions (the only dimension considered in this example is time) containing information that can be used to distinguish between the components. If the number of dimensions one wishes to analyse becomes large, then it may be appropriate for the analyst to investigate the use of a multivariate algorithm to distinguish between signal and background samples of events. Chapter 10 introduces a number of algorithms that can be used for such problems.

1.4 Summary

Students encountering statistical methods in an undergraduate laboratory course often express a lack of enthusiasm for the relevance of the topic. In an attempt to illustrate the motivation for this sometimes dry subject, this chapter has introduced three common problems encountered by students studying physics. There are different levels of sophistication in the way that the data are analysed for each of these problems. Rudimentary measurements made, sometimes even with basic apparatus, can often be improved upon with a little thought spent planning how to use the equipment and the application of a more refined knowledge of statistics. More advanced techniques can be used in order to extract quantitative information from shapes of distributions of data, assuming some known or presumed underlying physical model. A number of forward references have been made in this chapter in order to guide the interested reader to the appropriate parts later in this book where one can find detailed discussions of the relevant concepts or techniques that have been applied. It is hoped that the reader will have gained an appreciation of the need to understand statistical methods in order to effectively analyse data through the examples discussed.

2
Sets

Before embarking upon a detailed discussion of statistics, it is useful to introduce some notation to help describe data. This section introduces elementary set theory notation and Venn diagrams.

The notion of a **set** is a collection of objects or elements. This collection can also be referred to as **data**, and the individual elements in the data can themselves be referred to as data (in the singular sense), or as an **event** or **element**. We usually denote a set with a capital letter, for example Ω. The element of a set is denoted by a lower case letter, for example either ω or ω_i, where the latter explicitly references the ith element of the set. The **elements of a set** are written within curly braces '{' and '}'. For example we can write a set Ω_{Binary} that contains the elements 1 and 0 as

$$\Omega_{Binary} = \{1, 0\}. \tag{2.1}$$

This is called the binary set, as it contains the elements required to represent a binary system. The order of elements in a set is irrelevant, so we could write Ω_{Binary} in an equivalent form as

$$\Omega_{Binary} = \{0, 1\}. \tag{2.2}$$

If we want to express the information that a given element is or is not a part of a set, we use the symbols $\in$ and $\notin$, respectively. For example we may write

$$0 \in \Omega_{Binary}, \text{ and } 1 \in \Omega_{Binary}, \tag{2.3}$$
$$\text{but } 2 \notin \Omega_{Binary}, \tag{2.4}$$

to express that both 0 and 1 are elements of Ω_{Binary}, but 2 is not an element of this set.

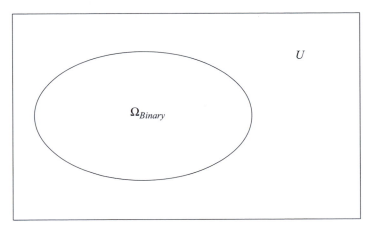

Figure 2.1 A Venn diagram illustrating the binary set Ω_{binary} and the universal set U.

Other useful sets include

$\mathbb{R}$ the set of all real numbers,

$\mathbb{R}^{\pm}$ the set of all positive/negative real numbers,

$\mathbb{C}$ the set of all complex numbers,

$\mathbb{Z}$ the set of all integer numbers (positive, negative, and zero),

$\mathbb{N}^{\pm}$ the set of all positive/negative integers ($0 \notin N^{\pm}$),

$\mathbb{Q}$ the set of all rational numbers ($p/q \in \mathbb{Z}$, where $p, q \in \mathbb{Z}$, and $q \neq 0$),

$\emptyset$ an empty set (one that contains no elements).

One can define a set in two ways (i) using a list, or (ii) using a rule. An example of defining a set using a list is the binary set Ω_{Binary} introduced above in Eq. (2.1). We can consider using a rule to define more complicated sets, for example

$$A = \{x | x \in \mathbb{R}^{+}, \text{ and } x > 2\}, \tag{2.5}$$

where A is the set of real positive numbers greater than 2. Another way to indicate a set of real numbers in some range is to use the following notation: $A \in [2, +\infty]$, which defines the set A to have the same elements as given in Eq. (2.5).

One other set that can be useful is the universal set, which contains all elements of interest, and is often denoted by U. It can be useful to graphically interpret sets in terms of a Venn diagram. Figure 2.1 shows the binary and universal sets represented in this way.

2.1 Relationships between sets

Now that elementary properties of sets have been introduced, it is useful to examine relationships between different sets. The following sections introduce additional operators that are relevant.

2.1.1 Equivalence

Two sets are said to be equal or equivalent if they contain the same elements as each other and nothing more. As in the case of the binary set described above as $\{0, 1\}$ or $\{1, 0\}$. For example if we consider $A = \{0, 1, 2, 3\}$ and $B = \{3, 2, 0, 1\}$, we can see that all of the elements of A occur in B and similarly all of the elements in B occur in A. So in this case $A = B$. In terms of a Venn diagram, A and B would appear as two completely overlapping regions.

2.1.2 Subset

Given a large set of data A, it can be useful to identify groups of elements from the set with certain properties. If the selection of identified elements form the set B, then B is called a subset of A and we may write

$$B \subset A, \qquad (2.6)$$

where the symbol $\subset$ denotes a subset.

For example if we consider the set $\mathbb{R}$, then this completely contains the set $\mathbb{R}^+$. We can consider $\mathbb{R}^+$ a subset of $\mathbb{R}$. This can be written as

$$\mathbb{R}^+ \subset \mathbb{R}. \qquad (2.7)$$

This notation can be extended to include the possibility that the two sets may be equivalent by replacing $\subset$ with $\subseteq$. Using this notation it follows that one could write $\mathbb{R}^+ \subseteq \mathbb{R}^+$ instead of $\mathbb{R}^+ = \mathbb{R}^+$ as the set $\mathbb{R}^+$ contains all of the elements necessary to define itself. However, this does in general introduce the ambiguity that the two sets are not necessarily equal. The symbol $\subseteq$ is the set notation analog of the inequality $\leq$. Figure 2.2 is a Venn diagram showing two sets A and B, in addition to the universal set. In this example $A \subset B$.

2.1.3 Superset

We can consider a related problem to that discussed above, where the set A is a subset of B, that is B contains all elements of A. In this case we call B the superset of A and may write this as

$$B \supset A, \qquad (2.8)$$

where the symbol $\supset$ denotes superset.

For example if we consider the set $\mathbb{R}^+$ we note that this is completely contained in set $\mathbb{R}$. So we can consider $\mathbb{R}$ a superset of $\mathbb{R}^+$, and we may write

$$\mathbb{R} \supset \mathbb{R}^+. \qquad (2.9)$$

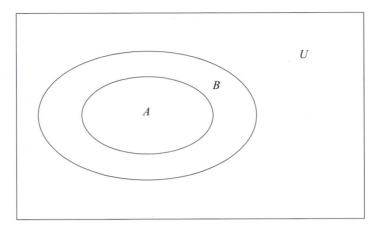

Figure 2.2 A Venn diagram illustrating two sets A and B where $A \subset B$, and the universal set U.

2.1.4 Intersection

Consider two sets A and B each containing a number of elements. If some of the elements contained in A also appear in B, then we can identify those common elements as the intersection of A and B. We may write

$$A \cap B = \{x | x \in A \wedge x \in B\}, \qquad (2.10)$$

where the symbol $\cap$ denotes intersection, and $\wedge$ denotes a logical and. If the intersection between sets A and B is an empty set we can write $A \cap B = \emptyset$, see Figure 2.3.

2.1.5 Set difference

Given the set A and the set B, it can be useful to identify the elements that only exist in A. The notation for the set of elements that exist in A, but not B is $A \setminus B$. For example if $A = \{1, 2, 3\}$, and $B = \{3, 4, 5\}$, then $A \setminus B = \{1, 2\}$. Similarly the elements of B that only exist in B are given by $B \setminus A = \{4, 5\}$.

2.1.6 Union

Consider two sets A and B each containing a number of elements. We may write the union of the two sets, which is a combination of all elements from both of the original sets, as

$$A \cup B = \{x | x \in A \vee x \in B\}, \qquad (2.11)$$

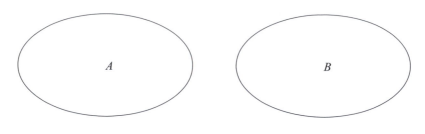

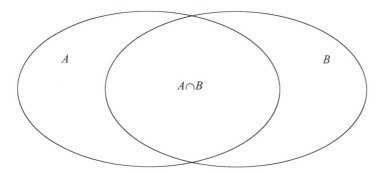

Figure 2.3 A Venn diagram illustrating two sets A and B where (top) $A \cap B = \emptyset$, and (bottom) $A \cap B$ is non-trivial. The universal set has been suppressed here.

where the symbol $\cup$ denotes union and $\vee$ denotes a logical or. This definition of $A \cup B$ implicitly includes the subset of elements $A \cap B$. If one considers the second case in Figure 2.3, it is clear that $A \cup B = A \cap B + A \setminus B + B \setminus A$.

2.1.7 Complement

The complement of some set A is the set of interesting elements not contained within A itself, and this is denoted by placing a bar above the set name. For example, the complement of A is $\overline{A}$. It follows that

- $A \cup \overline{A} = U$ as together A and its complement contain all of the interesting elements,
- $A \cap \overline{A} = \emptyset$, as the complement of a set and the original set do not have any elements in common,
- $\overline{\overline{A}} = A$, as the complement of $\overline{A}$ is the complement of the complement of the set A, which is the original set A.

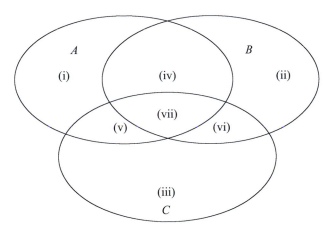

Figure 2.4 A Venn diagram illustrating three partially overlapping sets A, B, and C.

Example. One can consider the situation where there are three partially overlapping sets A, B, and C, as shown in Figure 2.4. Here the universal set has not been explicitly shown. In this scenario there are seven distinct regions.

Region	Elements
(i)	Only existing in set A: $A \setminus B \setminus C = \{x \mid x \in A \land (x \notin B \land x \notin C)\}$
(ii)	Only existing in set B: $B \setminus A \setminus C = \{x \mid x \in B \land (x \notin A \land x \notin C)\}$
(iii)	Only existing in set C: $C \setminus A \setminus B = \{x \mid x \in C \land (x \notin A \land x \notin B)\}$
(iv)	$A \cap B$ (this includes regions iv and vii)
(v)	$A \cap C$ (this includes regions v and vii)
(vi)	$B \cap C$ (this includes regions vi and vii)
(vii)	$A \cap B \cap C$

2.2 Summary

This section serves as a short summary of some of the rules discussed in this chapter, and where appropriate extends this to introduce more complicated notation that can be useful when manipulating sets.

(1) For a set A

$$A \cap A = A, \tag{2.12}$$

$$A \cup A = A, \tag{2.13}$$

$$A \setminus A = \emptyset, \tag{2.14}$$

where $\cap$ (intersection) and $\cup$ (union) operators are commutative.

(2) For two sets A and B it follows that

$$A \cap B = B \cap A, \tag{2.15}$$
$$A \cup B = B \cup A. \tag{2.16}$$

(3) Similarly these operators obey associativity which follows on naturally from their commutative nature

$$(A \cap B) \cap C = A \cap (B \cap C) = (A \cap C) \cap B \ldots \tag{2.17}$$
$$(A \cup B) \cup C = A \cup (B \cup C) = (A \cup C) \cup B \ldots \tag{2.18}$$

(4) These operators also follow a distributive law, so

$$A \cap (B \cup C) = (A \cap B) \cup (A \cap C), \tag{2.19}$$
$$A \cup (B \cap C) = (A \cup B) \cap (A \cup C). \tag{2.20}$$

(5) Two ways of defining an identity relation for a set A are

$$A \cup \emptyset, \tag{2.21}$$
$$A \cap U. \tag{2.22}$$

(6) The notion of a set, and the syntax provided by set algebra introduced in this section provides a framework that can be used to simplify the discussion of variables and data. Familiarity with the notation introduced in this section is assumed when discussing data samples and techniques to analyse data in the following sections.

Exercises

2.1 Given the sets $A = \{1, 2, 3, 4\}$ and $B = \{1, 3, 5, 9\}$, where these form the universal set, compute the following; (i) $A \cup B$, (ii) $A \cap B$, (iii) $A \setminus B$, (iv) $B \setminus A$, and (v) $\overline{A}$, and $\overline{B}$.

2.2 Given the sets $A = \{0, 2, 5, 6\}$ and $B = \{1, 3, 5, 9\}$, compute the following; (i) $A \cup B$, (ii) $A \cap B$, (iii) $A \setminus B$, and (iv) $B \setminus A$.

2.3 Draw a Venn diagram to illustrate the relationships between two sets A and B for the following cases, noting where appropriate $A \setminus B$, $B \setminus A$, $A \cap B$, $A \cup B$: (i) $A \cap B = \emptyset$ and (ii) $A \cup B = A$, where $A \neq B$.

2.4 If set $A = \mathbb{R}$, and set B contains the elements 1, 2, 3, 4, 5, what is $A \cup B$?

2.5 If set $A = \mathbb{R}^+$ and set $B = \{0, \mathbb{R}^-\}$, what is $A \cup B$?

2.6 If set $A = \mathbb{Z}$, and set B contains the elements 1, 2, 3, 4, 5, what is $A \cap B$?

2.7 If set $A = \mathbb{R}^+$ and set $B = \mathbb{R}^-$, what is $A \cap B$?

2.8 If $A = \mathbb{C}$, set $B = \mathbb{Z}$, and set $C = \mathbb{R}^+$, what is $A \cup B \cap C$?

2.9 If $A = \mathbb{C}$, set $B = \mathbb{Z}$, and set $C = \mathbb{R}^+$, what is $A \cup B \cap \overline{C}$?

2.10 Write down the decimal set Ω_{Decimal}.

2.11 Write down the set of primary colours.

2.12 What is $\mathbb{C} \cup \mathbb{Z}$?

2.13 What is $\mathbb{Z} \cap \mathbb{R}$?

2.14 What is $\{x | x > 5\} \cap \{x | x < 10\}$?

2.15 What is $\{x | x > 5\} \cup \{x | x < 10\}$?

The definition of Bayesian probability depends on a subjective input often based on theory. In many circumstances, the theory can be relatively advanced, based on previous information that is relevant for a particular problem, however, the opposite is also often the case and the theory expectations can be extremely crude. ***Bayes' theorem*** states that for data A given some theory B

$$P(B|A) = \frac{P(A|B)}{P(A)} P(B). \tag{3.4}$$

The terms used in this theorem have the following meanings.

$P(B|A)$ is called the ***posterior probability*** and it represents the probability of the theory or hypothesis B given the data A.

$P(A|B)$ is the probability of observing the data A given the theory or hypothesis B. This is sometimes referred to as the ***a priori probability***.

$P(B)$ is called the ***prior probability***. This is the subjective part of the Bayesian approach and it represents our degree of belief in a given theory or hypothesis B before any measurements are made.

$P(A)$ the probability of obtaining the data A, this is a normalisation constant to ensure that the total probability for anything happening is unity. More generally $P(A)$ is a sum over possible outcomes or hypotheses B_i, i.e. $P(A) = \sum_i P(A|B_i)P(B_i)$.

Sometimes it is not possible to compute the normalisation constant $P(A)$, in which case it is possible to work with the following proportionality derived from Eq. (3.4)

$$P(B|A) \propto P(A|B)P(B). \tag{3.5}$$

The use of this proportionality is discussed further in Section 8.4.

The notation introduced here naturally lends itself to discrete problems, however, one can extend the notation to accommodate continuous variables. In this case the quantities $P(B|A)$, $P(A|B)$, $P(B)$, and hence $P(A)$ become functions of the continuous variables, and may also become functions of some parameters required to define the shapes of the underlying hypotheses being tested. Several examples of the use of Bayes' theorem to compute posterior probabilities can be found in Section 3.7.

3.2.1 Priors

The subjective part of the Bayesian prescription of probability is the choice of the prior probability. What form should this take? If the possible outcomes are discrete then the prior probability $P(B_i)$ will correspond to some number. If, however, the prior probability depends on the value of some continuous variable x, then the prior probability will take the form of a distribution $P(B_i) = \theta(x)$. Instead of assigning a single number for the prior one assumes the functional form $\theta(x)$ of a prior probability distribution. Often it is convenient to choose a **uniform prior** probability distribution (see Appendix B.2.3), so effectively taking $P(B) = constant$. Such a choice is a representation of our ignorance of a problem where all values of x are taken to be equally likely, however, this is not always well motivated. One constraint on the choice of prior is that the result should be stable, and independent of this choice. Therefore one can check to see how a result depends on a given prior in order to establish the reliability of an estimate.

When dealing with a discrete problem, for example 'Will it rain tomorrow?' from Section 3.7.4, the prior probability that it will be rain is a discrete quantity associated with the hypothesis that it will rain, or the complement, that it will not rain. In order to compute a value to associate with the prior, one needs additional information, for example that it typically rains 131 days of the year in England, thus the prior probability that it will rain can be computed for this geographical location: $P(rain) = 131/365$. The posterior probability obtained for this example is an 88.7% chance of rain.

Without that prior information, one might assume that it would be equally likely to rain or not, and the choice of $P(rain)$ could be different. It would not be unreasonable to assume $P(rain) = 0.5$ in such a scenario. If one repeated the calculation in Section 3.7.4 with $P(rain) = 0.5$, and $P(no\,rain) = 0.5$, then the probability obtained for rain tomorrow is 93.3%. While this numerically differs from the original calculation, one can see that the overall conclusion remains unchanged. It is more likely to rain than not, given that rain has been forecast.

When dealing with a continuous problem there are an infinite number of possible priors $\theta(x)$ to choose from; however, in practice one may prefer to investigate results using a limited number of priors based on any available information. As mentioned above a common choice is a uniform prior

$$\theta(x) = \text{constant}, \tag{3.6}$$

where the prior is normalised appropriately to preserve the total probability. There is no variable dependence for a uniform prior, and the use of this can simplify computation (for example see the upper limit calculation discussed in Section 9.7.3). In the total absence of information a uniform prior over all allowed physical values of the problem space may be reasonable. However, if results already exist that may provide a weak constraint on the value of an observable then one could incorporate that information in the prior.

A detailed discussion of priors is beyond the scope of this book, and the interested reader is encouraged to refer to the references listed at the end of this book (in

Table 3.1 *The probabilities of rolling two dice and obtaining the outcomes shown for the first die (rows) and the second die (columns).*

	1	2	3	4	5	6
1	1/36	1/36	1/36	1/36	1/36	1/36
2	1/36	1/36	1/36	1/36	1/36	1/36
3	1/36	1/36	1/36	1/36	1/36	1/36
4	1/36	1/36	1/36	1/36	1/36	1/36
5	1/36	1/36	1/36	1/36	1/36	1/36
6	1/36	1/36	1/36	1/36	1/36	1/36

particular, James, 2007; Silvia and Skilling, 2008). The important issue to note has already been mentioned above: *any physical result reported should be independent of the choice of prior*, hence in general one should check a given result by validating that it is stable with respect to the choice of prior. If a result or conclusion depends strongly on the prior then this may indicate insufficient data is available to draw a conclusion, or poor choice of prior, or both. In general one should be careful to include sufficient information used to derive the result so that a third party can understand exactly what has been done.

3.3 Classic approach

The classic approach to probability is restricted to situations where it is possible to definitely assign a probability based on existing information. For example, if one rolls an unbiased (fair) six-sided die, then the probability of having the die land face up on any particular side is equal to $1/6$. The classic approach to probability can be used to calculate the probability of many different outcomes including the results of throwing dice, card games and lotteries. However, this approach is limited to situations where the outcomes are finite and well defined.

If one considers the scenario of rolling two fair dice, it is possible to construct a matrix of possible outcomes for the first die and for the second. The probabilities for the set of outcomes is illustrated in Table 3.1, where each specific outcome has equal probability of occurring, namely $1/6 \times 1/6 = 1/36$. Possible outcomes of interest can be grouped together. For example, the probability of rolling two sixes is $1/36$ as this is one distinct outcome, which is the same as that for any other single outcome. If one cares only about rolling a double (both die giving the same value), then the probability is the sum of probabilities for the combinations 11, 22, 33, 44, 55, and 66, which is $1/6$. If one cares only that a particular configuration of two different numbers is the result, for example that a one and a three is given, then there are two possible combinations to obtain that solution, hence the probability is 2/36.

3.4 Frequentist probability

The frequentist approach interprets probability in terms of the relative frequency of a particular repeatable event to occur. A consequence of this is that one has to be able to discuss a repeatable event or experiment in order to apply frequentist techniques.

Consider a repeatable experiment where the set of outcomes is described completely in Ω. If there is a subset of n interesting events I, so $I \subset \Omega$, and a total of N elements in Ω, then the *frequentist probability* of obtaining an interesting event is given by

$$P(I) = \lim_{N \to \infty} \left(\frac{n}{N} \right). \tag{3.7}$$

At first glance it may appear that it is impractical to compute a frequentist probability as we need to have an infinite set in order to produce an exact result. In practice, however, it turns out that in many cases one can compute an exact value of $P(I)$ without resorting to an infinite set using the classic approach. In other cases it is possible to compute $P(I)$ to sufficient precision with finite N. At all steps, frequentist probability is computed by following a defined set of logical steps. This in itself is an attractive feature of the prescription for many people. This approach is not without problems, and to illustrate this the important issue of checking the coverage quoted for an interval or limit computed using a frequentist approach is examined in Section 7.8.3.

3.4.1 Which approach should I use?

Sometimes individuals become very attached to one of the main schools of thought and lose focus on the problem at hand in order to concentrate on which statistical philosophy is correct. While the classic or frequentist approach can lead to a well defined probability for a given situation, it is not always usable. In such circumstances one is then left with only one option: use Bayesian statistics.[1] In practice these approaches can give somewhat different predictions when the data are scarce to come by; however, given sufficiently large samples of data the probabilities computed using one method are much the same as those computed using the other. For this reason the choice between approaches should be considered arbitrary as

---

[1] For example if it is not possible to repeat an experiment, then the computation of probability lends itself to the Bayesian approach.

long as both approaches remain valid. In fact it can be very useful to consider the results of using both approaches when trying to interpret data, where the results obtained can be considered as cross-checks of each other. The more complex an analysis, the more desirable it may be to perform such an independent cross-check to gain confidence that there are no significant problems in the analysis.[2]

3.5 Probability density functions

A *probability density function* (PDF) is a distribution where the total area is unity and the variation of the PDF is related to the probability of something occurring at that point in the parameter space. If we consider the simple case of a one-dimensional PDF that describes a uniform distribution of some variable x between the values -1 and $+1$, then the PDF $f(x)$ is simply a straight line with zero gradient. The function itself is given by

$$f(x) = \frac{1}{A}, \tag{3.8}$$

where A is a *normalisation constant* determined by

$$A = \int_{-1}^{1} dx, \tag{3.9}$$

$$= 2. \tag{3.10}$$

So in this instance, the PDF is given by

$$f(x) = \frac{1}{2}. \tag{3.11}$$

If a set of data is described by a PDF, then we can use the PDF to tell us if we are more likely to find a data point at $x = x_1$ than at a neighbouring point $x = x_2$.

For example, if we consider the uniform PDF above, then it follows that we are equally likely to find data points anywhere between the minimum and maximum values of x. We have already encountered this function above in the context of a uniform prior, where here the valid domain is $x \in [-1, +1]$.

More generally a PDF will be described by some function $f(x)$, where

$$\int_{a}^{b} f(x)dx = 1, \tag{3.12}$$

---

[2] This is especially true when it comes to algorithms dependent on computer programs that may easily incorporate coding errors. If two independent methods provide consistent results one can be confident that either no such errors exist or are small, or alternatively complementary mistakes have been made. The latter possibility, while an option, is usually very unlikely.

and a and b represent the limits of the valid domain for the function. The probability of obtaining a result between x and $x + dx$ is $f(x)dx$.

PDFs are useful concepts in many branches of statistical data analysis, and in particular they are useful in defining models that can be used to extract detailed information about data through optimisation techniques such as those described in Chapter 9. Many commonly used PDFs can be found in Chapter 5 and Appendix B. While we often consider PDFs to be continuous distributions, it can also be useful to use discrete distributions to represent a PDF.

3.6 Likelihood

The **likelihood** $\mathcal{L}$ of something to occur is proportional to the probability and has the property

$$P = C \cdot \mathcal{L}, \tag{3.13}$$

where C is some constant of proportionality. The function $\mathcal{L}$ is typically normalised such that the maximum value is unity. For example, if one considers Eq. (3.11), then the corresponding likelihood function is given by

$$f(x) = 1, \tag{3.14}$$

where the constant of proportionality between the PDF and likelihood is $C = 1/2$. In addition to using the likelihood function $\mathcal{L}$, one often uses $-\ln \mathcal{L}$ in order to take advantage of the computational benefit associated with logarithms converting products into summations.

Whereas probability gives one the ability to discuss the absolute expectation of some experiment, the use of likelihood provides a natural way to investigate the relative expectations of an outcome. A number of practical applications of likelihood include the constructs based on ratios of likelihoods or minimising $-\ln \mathcal{L}$ in the case of the maximum likelihood fit approach discussed in Chapter 9. A detailed treatment of likelihood can be found in Edwards (1992).

3.7 Case studies

This section discusses the outcomes of some familiar events that depend on chance, including the outcomes of tossing a coin, a lottery, and a popular card game. The final two examples demonstrate the use of a Bayesian approach. These are examples of what is more generally referred to as Game Theory.

3.7.1 Tossing a coin (classical)

We can perform the repeatable experiment of tossing a coin. Each coin toss is an independent event and has two possible outcomes. We denote these outcomes corresponding to the coin landing heads-up and tails-up with H and T, respectively. If the coin is unbiased there is an equal probability of the coin toss resulting in H or T. If we run an ensemble of 20 experiments with an unbiased coin, then we will obtain a random sequence of results. One such set of data is the sequence $THHTHHTTTTTTHTTHHHHH$. There are $10H$ and $10T$ in this outcome, which is consistent with our naive expectations that on average half of the time we will obtain the result H and half of the time we will obtain T.

Is it possible to toss the coin 20 times and obtain the result $20H$? Yes of course, this is just one of the valid outcomes. The probability of obtaining H for any particular event is $1/2$, hence the probability of obtaining T is also $1/2$. So the probability of obtaining $20H$ is $(1/2)^{20} = 1/1\,048\,576$. Now consider the particular result we obtained above: $THHTHHTTTTTTHTT\,HHHHH$. The probability for this to happen is also given by $(1/2)^{20} = 1/1\,048\,576$. We can express this result in the following way: if you do something a million times, then expect an event with probability of a million to one to happen. Our particular outcome of $10H10T$ has a probability of a million to one, but so does $20H$ (and for that matter $20T$). If we don't care about the ordering of H and T, then there are many ways to toss a coin in order to obtain a $10H10T$ outcome where each particular outcome has a probability of $(1/2)^{20}$. In fact the number of distinct $10H + 10T$ outcomes from flipping a coin 20 times is given by $20!/10!10!$. There are $184\,756$ combinations that result in a $10T + 10H$ result, each with a probability of $(1/2)^{20}$. So the probability of tossing a coin 20 times and obtaining one of these solutions is 17.6%. A formal treatment of this problem is discussed in the context of the binomial distribution in Section 5.2.

3.7.2 The national lottery (classical)

Many countries run national or regional lotteries. The UK's National Lottery is an example of a unbiased experiment with many possible outcomes. This is more complicated than tossing a coin, as six numbers are picked at random from a total of 49 numbers in the sequence 1 through 49. Each number can be picked only once, so after the first number is picked, there are 48 remaining numbers to choose from, and so on. If the six numbers chosen by the lottery machine match the numbers on a lottery ticket purchased by a member of the public they will win the jackpot. But what is the probability of this event occurring?

In order to determine this we need to be able to calculate the number of possible outcomes, and the number of possible successful outcomes for each lottery draw. There are 49! combinations of numbers. The order with which the winning numbers are selected does not matter, so there are 6! ways of selecting these six numbers. The remaining problem to solve is to determine how many combinations will result in the wrong set of numbers being selected. In selecting six numbers, we leave a further 43 numbers unselected in a complementary set. The total number of possible combinations of this complementary set is 43!. We can compute the probability of selecting the six winning numbers as the ratio of possible winning combinations divided by the total number of combinations. As the total number of winning combinations is given by the product of the combinations of the winning six numbers and the complementary set, the ratio is:

$$\frac{6!43!}{49!} = \frac{6!}{49 \times 48 \times 47 \times 46 \times 45 \times 44} = \frac{1}{13\,983\,816}. \tag{3.15}$$

So the probability of winning the jackpot is 1 in 13 983 816. The notation $49!/(6!43!)$ is often written in short hand as $^{49}C_6$. The quantity $^{49}C_6$ corresponds to the number of unique combinations of selecting six numbers, leaving a set of 43 complementary numbers from a total set of 49 numbers. As successive lottery draws are independent, the probability calculated here is the probability of winning the lottery in any given draw. Given this, we can see that one needs to play the national lottery 13 983 816 times using unique combinations of numbers in any given draw in order to be certain of winning the jackpot.

3.7.3 Blackjack (classical)

Blackjack is a card game, where the aim is to obtain a card score of 21 or less, being as close to that number as possible, with as few cards as possible. In this game the value of numbered cards is given by the number on the card, picture cards count for 10 points, and aces can count for either 11 points or 1 point. So in order to win the game with 21 points and two cards dealt, the player must have a picture card or 10, and an ace. There are 16 cards with a value of 10 points, and four cards with a value of 11 points out of a pack of 52 cards. Lets consider this simplest case of being dealt an ace and a 10 point card. The probability of being dealt one ace is four out of 52 times, or 7.7%. The probability of subsequently being dealt a 10 point card is 16 out of 51 times, or 31.4% (assuming that there are no other players). So the combined probability of being dealt an ace followed by a 10 point card is the product of these two probabilities: 2.4%. We don't care if we are dealt the ace as the first or second card, so we can also consider the possibility that we are dealt the 10 point card before the ace. This probability is also 2.4%. So the total probability of being dealt an ace and a 10 point card is the sum of these two probabilities, namely 4.8%. In reality the game is played by more than one person, so the probabilities are slightly modified to account for the fact that more than two cards are dealt from the pack before any of the players can reach a total point score of 21. With a little thought it is not difficult to understand how likely you are to be dealt any given combination of cards at a game such as this one.

3.7.4 Will it rain tomorrow? (Bayesian)

A weather forecast predicts that it will rain tomorrow. When it rains, 70% of the time, the forecast was correctly predicting rain, and when there was no rain forecast, only 5% of the time would it rain. It rains on average 131 days of the year in England. So, given this information, what is the probability that it will rain tomorrow given the forecast? Starting from Bayes theorem,

$$P(B|A) = \frac{P(A|B)}{P(A)} P(B), \tag{3.16}$$

where the hypothesis that it will rain is given by $B = rain\ forecast$, and the data of it actually raining tomorrow is A. We know that the prior probability of rain, $P(rain)$, is $131/365 = 0.3589$, and the corresponding prior probability that it will not rain is $P(no\ rain) = 234/365 = 0.6411$. We are told above that $P(rain|rain\ forecast) = 0.7$, and that $P(rain|not\ forecast) = 0.05$, so the normalisation $P(A)$ is given by

$$P(A) = P(rain|rain\ forecast)P(rain) +$$
$$P(rain|not\ forecast)P(no\ rain) \tag{3.17}$$
$$= 0.7 \times 0.3589 + 0.05 \times 0.6411, \tag{3.18}$$

and hence the posterior probability is

$$P(rain\ forecast|rain) = \frac{0.7 \times 0.3589}{0.7 \times 0.3589 + 0.05 \times 0.6411}. \tag{3.19}$$

Thus the probability that it will rain tomorrow given the forecast is 88.7%, so it may be a good idea to pick up an umbrella given this forecast.

3.7.5 The three cups problem (Bayesian)

Consider the case when someone presents three cups C_1, C_2, and C_3, only one of which contains a ball. You're asked to guess which cup contains the ball, and on guessing you will make a measurement to identify if you have found a ball or not. What is the probability of finding a ball under cup C_1? In this example we will use

Bayes' theorem to compute the probability, so we start from:

$$P(B_1|A) = \frac{P(A|B_1)}{P(A)}P(B_1). \tag{3.20}$$

Here hypothesis B_1 corresponds to the ball being being found under cup C_1. Intuitively we can consider that all cups are equal, and so one can assume that the probability of a cup containing a ball would be $1/3$. That means that there is a probability of $1/3$ that cup C_1 will contain a ball. This gives us the prior probability of $P(B_1)$ based on our belief that the cups are all equivalent or unbiased. The prior probabilities that the ball will be found under C_2 or C_3, given by $P(B_2)$ and $P(B_3)$, respectively, are also $1/3$.

So we have determined the prior probabilities, we now need to determine the probability of obtaining the data $P(A)$, and the probability of observing the outcome A given the prior B_1. The probability of observing the ball under cup C_1 is $1/3$, which is $P(A|B_1)$, and similarly one finds the probabilities of $1/3$ each for observing the ball under C_2 or C_3. The only thing that is still to be calculated is $P(A)$ which in general is given by

$$P(A) = \sum_i P(A|B_i)P(B_i), \tag{3.21}$$

$$= P(A|B_1)P(B_1) + P(A|B_2)P(B_2) + P(A|B_3)P(B_3), \tag{3.22}$$

where the probability $P(B_i)$ is the prior that we find the ball under the ith cup, and B_i is the corresponding hypothesis (i.e. under which cup we look). So

$$P(A) = \left(\frac{1}{3} \times \frac{1}{3}\right) + \left(\frac{1}{3} \times \frac{1}{3}\right) + \left(\frac{1}{3} \times \frac{1}{3}\right), \tag{3.23}$$

$$= \frac{1}{3}, \tag{3.24}$$

as we consider all cups to be equal and thus $P(B_i) = 1/3$ for all i. So in this example the normalisation term given by $P(A) = 1/3$, the prior probability $P(B_1) = 1/3$, thus the final quantity to be determined is $P(A|B_1)$. This is the probability of obtaining an outcome A given the hypothesis B_1, which again is $1/3$. Now returning to Bayes' theorem we find

$$P(B_1|A) = \frac{1/3}{1/3} \times 1/3 = 1/3. \tag{3.25}$$

This is the same result as one would have determined using a classical approach.

3.8 Summary

A number of elementary concepts have been introduced in this section. The main points raised are as follows.

(1) The underlying axioms that are used to discuss probability include the fact that the probability for something to happen is bound between zero (an event will not occur), and one (an event is certain to occur), i.e. $0 \leq P \leq 1$. Corollaries of this are:
   - probability is not negative,
   - the total probability for all possible outcomes is one.

(2) If two independent events i and j may occur, then the probability for one of them to happen is the sum of the probabilities for each of the independent events to occur, i.e. $P_{i+j} = P_i + P_j$.

(3) If one event i occurs, and this is followed by the uncorrelated event j, then the probability for this to happen is the product of the probabilities $P_{ij} = P_i \cdot P_j$.

(4) The quantification of how likely the outcome of some event might be can be approached in different ways. The two main schools of thought on the subject are that of Bayesian inference, and that of frequentist determination. In some cases, the underlying assumptions of the approach will dictate which method(s) are applicable to a given problem, and when both Bayesian and frequentist approaches are valid, it becomes a matter of personal choice as to which approach to use with a particular problem.
   - Bayesian probability is computed using Bayes' theorem which is given by Eq. (3.4).
   - Frequentist probability is computed by taking the limiting case of an infinite number of measurements found in Eq. (3.7).
   - There are many situations where one can resort to using a classical approach to determine an exact probability without having to resort to performing a frequentist treatment using the limiting case of Eq. (3.7).
   - In the case of repeatable experiments both Bayesian and frequentist approaches are valid; however, if a measurement is not repeatable, then only the Bayesian approach remains an option.

(5) One can represent probability in the form of a function of some continuous variable x. Such a probability density function (PDF) is normalised to a total area of unity.

(6) The likelihood function is related to a PDF via Eq. (3.13) where the constant of proportionality is chosen in order to set the maximum value of the likelihood to one.

Exercises

3.1 Compute the probability of being dealt an ace from a new deck of cards.

3.2 Two people are playing blackjack. What is the probability that the first two cards being dealt are both aces?

3.3 Compute the probability of being dealt either a picture or a 10 point card from a new deck of cards.

3.4 In a game of blackjack with two participants, compute the probability of being dealt an ace from a deck of cards, followed by either a picture or a 10 point card, assuming that the first card is dealt to you.

3.5 The weather forecast tomorrow is for rain; 70% of the time when it rains, the rain has been correctly forecast. When there is no rain forecast, it rains 5% of the time, and on average it rains 30 days of the year. Given this, what is the probability that it will rain tomorrow?

3.6 A storm is forecast for tomorrow. When storms occur 95% of the time, the forecast was correct, and when there was no storm was forecast, only 2% of the time would one happen. Storms occur on average 20 days of the year in England. Given this, what is the probability that a storm will occur tomorrow?

3.7 The weather forecast tomorrow is for rain; 70% of the time when it rains, the rain has been correctly forecast. When there is no rain forecast, it rains 5% of the time, and on average it rains 100 days of the year. Given this, what is the probability that it will rain tomorrow? Compare the result obtained with that assuming that on average it is equally likely to rain or not.

3.8 The weather forecast tomorrow is for rain; 90% of the time when it rains, the rain has been correctly forecast. When there is no rain forecast, it rains 5% of the time, and on average it rains 30 days of the year. Given this, what is the probability that it will rain tomorrow? Compare the result obtained with that assuming that on average it is equally likely to rain or not.

3.9 If there are two doors leading from a corridor, where nothing but an empty room lies behind one, and a staircase lies behind the other, use Bayes' theorem to compute the probability of opening the left-hand door and finding an empty room.

3.10 You appear on a gameshow where the host presents you with three doors. Behind one of the doors there is a car and if you correctly select this door you will win it. There is nothing behind the other two. You select one of the doors. After this, the host opens one of the two remaining doors to reveal nothing behind it. What should you do to maximise your chance of winning the car, and what is the corresponding probability?[3]

3.11 The parabola $y = N(3 - x)^2$ corresponds to PDF for $|x| < 3$. Compute the normalisation constant N, and re-express the PDF as a likelihood function with a most probable value of one.

3.12 Compute the PDF normalisation coefficient N for the function given by $y = N(1 + x + x^2)$ in the range $x \in [0, 2]$, and write down the corresponding likelihood function with a most probable value of one.

3.13 Compute the PDF normalisation coefficient for an exponential distribution approximating the decay of a muon, with lifetime $\tau_\mu = 2.2\,\mu s$, over the range $t \in [0, 20]\,\mu s$, and write the corresponding PDF and the likelihood function, where $\max[\mathcal{L}] = 1$.

3.14 What is the likelihood ratio for $x = 1$ compared to $x = 2$ for the function $\mathcal{L} = e^{-x}$, where $x \in \mathbb{R}^+$?

3.15 What is the likelihood ratio for $x = -1$ compared to $x = +1$ for the function $\mathcal{L} = e^{-x^2}$?

---

[3] Note this is the famous Monty Hall problem, named after the gameshow host.

4

Visualising and quantifying the properties of data

4.1 Visual representation of data

4.1.1 Histograms

It is often very convenient to describe data in a succinct way, either through a visual representation or through a brief quantitative description. Given a sample of data, for example the results of an ensemble of coin toss experiments, such as those described in Chapter 3, it can be more instructive to view the results tabulated in a histogram (see Figure 4.1) of the data. A **histogram** is a graphical representation where data are placed in discrete bins. Normally the content of each bin is shown on the vertical axis, and the value of the bin can be read off of the horizontal axis. In this particular coin tossing experiment there were $27H$ (heads-up) and $23T$ (tails-up) obtained. The benefit of drawing a histogram over providing the corresponding data in a table is that one can quickly obtain an impression of the shape of the data as a function of the different bins as well as the relative bin content. The coin tossing experiment is simple and the benefits of using histograms becomes more apparent when considering more complex problems.

The integral of a histogram that contains some number of bins N and some number of entries (or events) M, is

$$A = \sum_{i=1}^{N} y_i w_i \Delta x_i,$$
(4.1)

where y_i is the the the number of entries, w_i is the entry weight, and Δx_i is the width of the ith bin. Hence the sum over y_i is M. Often the entry weight for each event is one and in such circumstances Eq. (4.1) can be simplified accordingly. The above summation implicitly notes that the bin width Δx_i may vary.

As histograms are examples of a **discrete distribution**, variables that are continuous are discretised when represented by a histogram. In doing this one does lose information, so there has to be a trade off between the width of bins and the

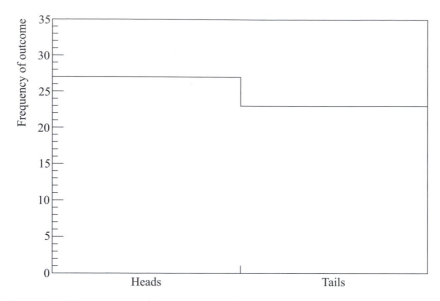

Figure 4.1 The outcome of an ensemble of coin tossing experiments resulting in either heads or tails.

number of entries in a given bin. Indeed the width of all of the bins need not be constant when trying to find a balance between bin content and bin width.

4.1.2 Graphs

We are often in the situation where we have a set of data points. These may represent estimates of some quantity y which vary as a function of x. In such situations, it can be beneficial to use a ***graph*** to represent these pairs of data in two dimensions as illustrated in Figure 4.2. Each point on the graph is drawn with an ***error bar***, indicating the uncertainty on the measured value of y. The concept of errors is discussed at length in Chapter 6.

It can be useful to overlay a curve or line on a graph to indicate trends; however, care should be taken to do so only when this is meaningful. There is no point in joining the dots between a set of data. When there is a trend, it is usually the result of fitting a model to the data that we want to overlay on a graph in order to compare the model to the data. This situation is discussed in Chapter 9.

4.1.3 Continuous distributions

If we consider the histogram shown in Figure 4.1, there are only two possible discrete outcomes for the data, and so it is only really sensible to bin the data as

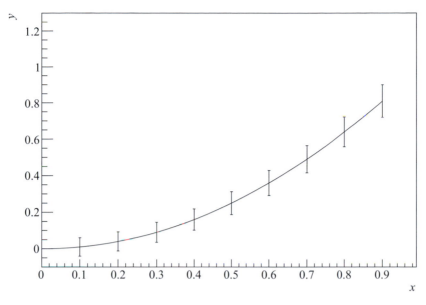

Figure 4.2 A graph of y versus x. An error bar in the y direction indicates the uncertainty on y.

shown. There are many circumstances encountered when the data are represented by a continuous outcome rather than a discrete one. In such cases information can be lost if the data are binned too coarsely. Taking the extreme case where the data are effectively infinite, then we can let the number of bins in a histogram become infinite. This limiting case results in a ***continuous distribution***, shown as the line in Figure 4.2. One can think of a continuous distribution for some data as the corresponding PDF re-normalised so that the integral is the total number of events found in the data sample. If there is an uncertainty associated with the distribution, this can be represented as a band in analogy with the error bars on a graph.

4.2 Mode, median, mean

If we take n repeated measurements of some observable x, it is useful to try and quantify our knowledge of the ensemble in terms of a single number to represent the value measured, and a second number to represent the spread of measurements. Thus, our knowledge of the observable x will in general include some central value of the observable, some measure of the spread of the observable, and the units that the observable is measured in. The spread of measurements is discussed in Section 4.3 below, here we discuss the representative value measured from an ensemble of data.

The ***mode*** of an ensemble of measurements is the most frequent value obtained. If the measurement in question is of a continuous variable, one has to bin the data in terms of a histogram in order to quantify the modal value of that distribution.

The ***median*** value of the ensemble is the value of x where there are an equal number of measurements above and below that point. If there is an odd number of measurements, then the point where there are exactly $(n-1)/2$ data above and below it is the median value. If there is an even number of measurements, then the median value is taken as the midpoint between the two most central values. The median value can be useful when it is necessary to rank data (for example in computing upper limits or some correlation coefficients, described later).

A better way to quantify the 'typical' value measured is to take an arithmetic average (or mean) of the individual measurements. The ***mean*** value is denoted either by $\bar{x}$ or $\langle x \rangle$ and is given by

$$\bar{x} = \langle x \rangle = \frac{1}{n} \sum_{i=1}^{n} x_i, \tag{4.2}$$

where x_i is the ith measurement of x. The mean value of a function $f(x)$ can be calculated in the same way using

$$\bar{f} = \frac{1}{n} \sum_{i=1}^{n} f(x_i). \tag{4.3}$$

If the function in question is a continuous one, then the average value of the function between $x = a$ and $x = b$ is given by the integral

$$\bar{f} = \frac{1}{b-a} \int_{x=a}^{b} f(x)dx. \tag{4.4}$$

It is possible to compute the average of a set of binned data; however, if rounding occurs in the binning process, then some information is lost and the resulting average will be less accurate than that obtained by using the above formulae.

Figure 4.3 shows the representation of a sample of data plotted in a histogram. This figure has two arrows to indicate the mode and mean. For this particular sample of data the mean is 5.1, and the mode is 6.5. The fact that the mode is greater than the mean is an indication that the data are asymmetric about the mean. We usually refer to such a distribution as being skewed, and in this case the data are skewed to the right. The skewness of a distribution is discussed further in Section 4.5.

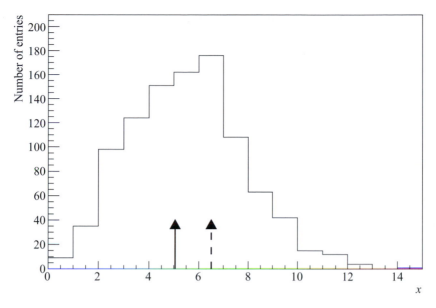

Figure 4.3 A sample of data represented by a histogram, with the mean (solid arrow) and mode (dashed arrow) indicated.

4.3 Quantifying the spread of data

4.3.1 Variance

The mean of an ensemble of data given by $\bar{x}$ doesn't provide any information as to how the data are distributed. So any description of a set of data just quoting a value for $\bar{x}$ is incomplete. We need a second number in order to quantify the spread or dispersion of data about the mean value. The average value of the deviations from the mean value is not a useful quantity as by definition this will be zero for a symmetrically distributed sample of data. We can consider the average value of the deviations from the mean squared as a measure of the spread of our ensemble of measurements, and this is called the **variance** $V(x)$, which is

$$V(x) = \frac{1}{n} \sum_{i=1}^{n} (x_i - \bar{x})^2, \tag{4.5}$$

$$= \frac{1}{n} \sum_{i=1}^{n} x_i^2 - 2x_i\bar{x} + \bar{x}^2, \tag{4.6}$$

$$= \frac{1}{n} \sum_{i=1}^{n} x_i^2 - \frac{1}{n} \sum_{i=1}^{n} 2x_i\bar{x} + \frac{1}{n} \sum_{i=1}^{n} \bar{x}^2, \tag{4.7}$$

$$= \overline{x^2} - 2\bar{x}^2 + \bar{x}^2, \tag{4.8}$$

$$= \overline{x^2} - \bar{x}^2. \tag{4.9}$$

So the variance is given by

$$V(x) = \overline{x^2} - \overline{x}^2. \tag{4.10}$$

The quantity $(x_i - \overline{x})$ is sometimes referred to as the **residual**[1] of x.

4.3.2 Standard deviation

The square root of the mean-squared (root-mean-squared or RMS) deviation is called the **standard deviation**, and this is given by

$$\sigma(x) = \sqrt{V(x)}, \tag{4.11}$$

$$= \sqrt{\overline{x^2} - \overline{x}^2}. \tag{4.12}$$

The standard deviation quantifies the amount by which it is reasonable for a measurement of x to differ from the mean value $\overline{x}$. In general we would expect to have 31.7% of measurements deviating from the mean value by more than 1σ, 4.5% of measurements to deviate by more than 2σ, and 0.3% of measurements to deviate by more than 3σ. If we performed a set of measurements where our results were more broadly distributed than this, we would worry about what might have gone wrong with the experiment. Chapters 6 and 8 discuss this issue in more detail.

The definition of the variance given in Eq. (4.5) is known to be biased for small n. The level of bias is given by a factor of $(n-1)/n$, so one often multiplies the variance given in Eq. (4.5) by the **Bessel correction** factor of $n/(n-1)$ in order to remove the bias on the variance; however, the standard deviation remains biased for small n even if the correction function is used. A brief discussion of how to determine the bias correction can be found in Kirkup and Frenkel (2006) and references therein. The unbiased variance is given by

$$V = \frac{1}{n-1} \sum_{i=1}^{n} (x_i - \overline{x})^2. \tag{4.13}$$

The corresponding form for the standard deviation is

$$\sigma(x) = \sqrt{\frac{1}{n-1} \sum_{i=1}^{n} (x_i - \overline{x})^2}. \tag{4.14}$$

It can be seen that both forms for $\sigma(x)$ are identical for $n \to \infty$, and it is better to use the second form for small samples of data. Scientists often prefer to use the

---

[1] The residual of an observable is the difference between the value given by a constraint equation or other system and a measurement. In this case it the residual is the difference between the ith measurement of an observable and the arithmetic mean determined from an ensemble of measurements of that observable.

standard deviation rather than variance when describing data as the former has the same units as the observable being measured. In contrast to this, mathematicians prefer to use the variance given in Eq. (4.13) as this is an unbiased statistic.

4.4 Presenting a measurement

A measurement consists of two pieces of information. Some quantification of the nominal value, and some indicator of how reliable that nominal value may be (the error). In general both the nominal value and the error will be determined in a certain unit. For example, consider the scenario where one has a piece of rope and is asked to determine how long it is. Given a ruler, it is possible to make a measurement of the rope length, and on doing this it is found to be 50 cm. It is extremely unlikely that the rope is exactly 50 cm long. The ruler will have some internal precision with which one may measure the position of a single point, for example if the scale is graduated in mm, then the precision to which one may determine a given point may be to within ± 0.5 mm. As there are two points to identify on the scale, and the same technique is used to determine both positions, we can assume that the total error on the measurement may be accurate to ± 1 mm. Now if we wish to discuss the length of the piece of rope as a measurement we may say that it is 50 ± 0.1 cm long. But if we do this, the nominal value and the estimate of how reliable that value is, are quoted to different precisions, which does not make sense. In one case we quote to the nearest centimetre, and in the other to the nearest millimetre. Conventionally we always choose to quote both numbers to the same precision. Would it make sense to quote this measurement as 50.0000 ± 0.1000 cm? No, the uncertainty represents the scale with which we can expect the true nominal value to lie within. Adding additional decimal places does not make the measurement any more precise, and indicates a lack of appreciation of a measurement. In this case it would be more appropriate to quote the length of rope as 50.0 ± 0.1 cm. The number of significant figures quoted is driven by the number of significant figures required to represent the uncertainty on the measurement. In Chapter 6 we will see that the average value of a set of measurements is given by $\langle x \rangle \pm \sigma_x$, where $\langle x \rangle$ is the arithmetic average and σ_x is the standard deviation of the data.

Just as there is a convention for the precision with which one quotes the nominal value and error of a measurement, there are conventions for the types of units: the so-called Système International (SI) units. For example the SI units for length, time, and mass are m (metre), s (second), and g (gramme). While it may seem to be pedantic that one specifies units when quoting measurements or tolerances on component designs, it is worth remembering why this is done. In order to compare results, or compare engineering designs and fabricate a working device to specifications, scientists and engineers from around the world need to be able to communicate with a single common language. When this does not happen confusion may arise. One such example of this is illustrated by the Mars Climate Orbiter spacecraft launched in 1998 (Mars Climate Orbiter, 1998). At a critical moment during the mission when the satellite was manoeuvring to enter orbit around Mars in 1999, the craft entered the atmosphere and burned up. The Mars Climate Orbiter Mishap Investigation Board concluded the source of the problem was a mismatch in the use of units: one software file expected inputs in metric units, while it was provided with data in imperial units. This simple mismatch resulted in the unfortunate loss of a spacecraft before it had a chance to begin the scientific part of its operational life. The Mars Climate Orbiter mission failed because the procedures in place to find problems, while correctly addressing many complicated issues, unfortunately overlooked the check that the units used were the same for an input file and a programme reading said file. When dealing with a complex problem one often finds that the most difficult issues are dealt with correctly, as they receive the most attention, while more straightforward problems occasionally get overlooked. From this extreme example we learn that there are three parts to a quantity or measurement being reported: the central value, the uncertainty, and the unit. If these are quoted then there is no ambiguity in the interpretation of the quantity. Hence the value, uncertainty, and unit are all important pieces of information to quote, and any result is incomplete if one of these is missing.[2] Often we are preoccupied with more complicated details when working on a problem and it is all too easy to overlook elementary details that could result in a simple but significant mistake being made. Great efforts are made in order to minimise such problems from arising, and the fact that such unfortunate mishaps are so rare is a testament to the ongoing efforts scientists and engineers make when designing new experiments.

If we now revisit the length of a piece of rope, it is clear that we have to quote the central value and uncertainty (50.0 ± 0.1) as well as the relevant units (cm) in order to avoid ambiguity. What might happen if we only quoted the value and uncertainty, without any units? A rock climber might naturally assume that the rope was 50.0 ± 0.1 m long as 50 m is a standard rope length. A person unfamiliar with that standard might correctly assume the scale was in centimetres, but there is an ambiguity that can not be resolved without additional information. To avoid mis-understandings we need to quote measurements not only in terms of central values and uncertainties, but also quoting these in well defined units. SI units and the guide to the expression of uncertainty in measurement (GUM) is a core topic

---

[2] There are a few exceptions to this statement, where the quantities being measured are themselves dimensionless by construction.

discussed in depth in the book by Kirkup and Frenkel (2006) for those readers wishing to explore this issue further.

4.4.1 Full width at half maximum

Sometimes instead of quantifying a distribution using the variance or standard deviation, scientists will quote the *full width at half maximum* (FWHM). This has the advantage that any extreme outliers of the distribution do not contribute to the quantification of the spread of data. As the name suggests, the FWHM is the width of the distribution (the spread above and below the mean) read off of a histogram of data at the points where the distribution falls to half of the maximum. The FWHM can be compared to the standard deviation of a Gaussian distribution by noting that

$$FWHM = 2.35\sigma, \tag{4.15}$$

however, some thought should be put into deciding if such a comparison is meaningful or sensible for a given set of data.

4.5 Skew

In the previous discussion, we formalised a way to quantify a distribution of data resulting from an ensemble of measurements with only two numbers: one number giving a single value describing the outcome of our measurement the *arithmetic mean*, and one number describing the spread of measurements the *standard deviation*. In the definition of standard deviation, we have implicitly assumed that the data are symmetric above and below the mean. This can be seen in Eq. (4.5) as we compute the sum of squared deviations from the mean. Our quantification of the data doesn't tell us about any possible asymmetry about the mean value. We can try and quantify such asymmetries using the *skew* or skewness of the data, which is a quantity derived from the third moment of x about its mean value and is given by

$$\gamma = \frac{1}{n\sigma^3} \sum_{i=1}^{n} (x_i - \overline{x})^3, \tag{4.16}$$

$$= \frac{1}{\sigma^3} \left(\overline{x^3} - 3\overline{x}\,\overline{x^2} + 2\overline{x}^3 \right). \tag{4.17}$$

The value of standard deviation used in the computation of γ should take into account the Bessel correction factor for small values of n where appropriate. However, one should also take care not to over analyse small samples (e.g. a few events) of data that could lead to misleading conclusions.

As with the standard deviation, there are alternate forms for the skew of a distribution. A simple form is given by Pearson's Skew, which is

$$\text{Skew} = \frac{mean - mode}{\sigma}. \tag{4.18}$$

This is zero by definition when the distribution of data is symmetric and the mean and mode coincide.

Skew can occasionally be useful. Quantities derived from higher-order moments exist, for example kurtosis is the corresponding variable constructed from the fourth moment, but they are not commonly used in physical sciences.

4.6 Measurements of more than one observable

4.6.1 Covariance

The previous discussion only considered measuring some observable quantity x, and how we can simply quantify an ensemble of n such measurements. Often we are faced with the more complicated problem of making measurements that depend on more than one observable. We can construct a variance between two observables x and y, commonly called the **covariance** $\text{cov}(x, y) = V_{xy} = \sigma_{xy}$ which is given by

$$\sigma_{xy} = \frac{1}{n} \sum_{i=1}^{n} (x_i - \overline{x})(y_i - \overline{y}), \tag{4.19}$$

$$= \overline{xy} - \overline{x}\,\overline{y}. \tag{4.20}$$

The usual caveat with regard to small n applies to the covariance.

If the values of x and y are independent, then the covariance will be zero. If, however, large values of $|x|$ tend to occur with large values of $|y|$, and similarly small values of $|x|$ with small values of $|y|$ then the covariance will be non-zero. We can extend our notation to an arbitrary number of dimensions, where we can denote the ith data point as $\underline{x_i} = (x_{(1)}, x_{(2)}, \ldots x_{(M)})_i$. Here the subscript i refers to the data set, and the subscripts with parentheses refer to the particular dimension of the data point. For example in a two-dimensional problem $x_{(1)} = x$ and $x_{(2)} = y$. Using this notation we can write an $M \times M$ **covariance matrix** (also called the **error matrix**) for an M-dimensional problem given by

$$V = \sum_{i,j} \sigma_{x_{(i)}x_{(j)}} = \sum_{i,j} \sigma_{ij}, \tag{4.21}$$

where i and j take values from one to M. The diagonals of this matrix, where $i = j$, are the variance of the ith observable. For example, if we consider the variables x

and y, the corresponding covariance matrix is

$$V = \begin{pmatrix} \sigma_x^2 & \sigma_{xy} \\ \sigma_{xy} & \sigma_y^2 \end{pmatrix}. \tag{4.22}$$

The covariance matrix appears to be a rather abstract entity; however, in Chapter 6 this matrix is revisited in the context of understanding how to combine uncertainties that are correlated, or indeed how to combine independent measurements of sets of correlated observables. In that more complete context the notion of the covariance between pairs of observables, and also the covariance matrix should make more sense.

4.6.2 Correlation

Covariance is a dimensional quantity, and it can be useful to work with a quantity that is a scale-invariant dimensionless measure of the dependence of one variable on another. The ***Pearson correlation coefficient*** ρ_{xy} (or just ρ without the subscripts indicating dimensions) is one such variable, where the covariance is normalised by the product of the standard deviations of x and y. This is given by

$$\rho_{xy} = \frac{\sigma_{xy}}{\sigma_x \sigma_y}, \tag{4.23}$$

$$= \frac{1}{n\sigma_x \sigma_y} \sum_{i=1}^{n} (x_i - \overline{x})(y_i - \overline{y}). \tag{4.24}$$

Again this can be generalised to an $M \times M$ matrix called the ***correlation matrix***. The elements of the correlation matrix are given by

$$\sum_{i,j} \rho_{ij} = \sum_{i,j} \frac{\text{cov}(x_{(i)} x_{(j)})}{\sigma_i \sigma_j}, \tag{4.25}$$

$$= \sum_{i,j} \frac{V_{ij}}{\sigma_i \sigma_j}, \tag{4.26}$$

$$= \sum_{i,j} \frac{\sigma_{ij}}{\sigma_i \sigma_j}, \tag{4.27}$$

where the diagonal elements ($i = j$) of this matrix are unity.

It can be seen from Eq. (4.23) that ρ_{xy} is zero if x and y are independent (following on from the previous discussion with respect to covariance). If x is completely dependent on y, then the possible values of ρ_{xy} are ± 1. In the case that x increases with increasing y $\rho_{xy} = +1$, and for the case where x decreases for increasing y $\rho_{xy} = -1$. Figure 4.4 illustrates some data samples with different

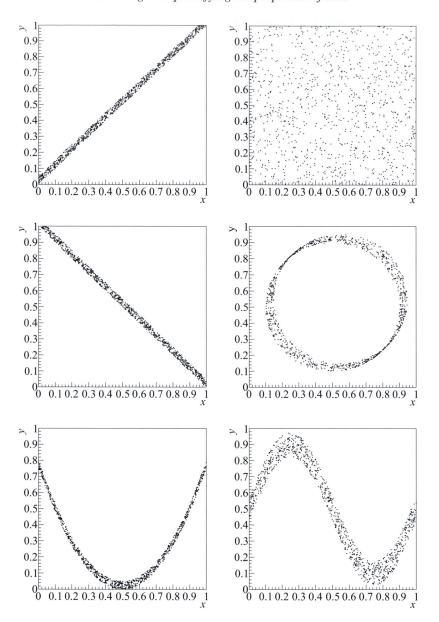

Figure 4.4 From left to right and top to bottom, different samples of data with Pearson correlation coefficients of $\rho_{xy} = +1, 0, -1$, and non-linear correlations with circular ($\rho_{xy} = 0$), quadratic ($\rho_{xy} = 0$) and sinusoidal ($\rho_{xy} = -0.77$) patterns.

correlation coefficients, the values of which are summarised in Table 4.1. For the linear or random samples, one can determine a reasonable estimate of the correlation coefficient by looking at the data. However, if the data have non-linear patterns, for example the circular or quadratic patterns shown in the figure, then the

Table 4.1 *Pearson and Spearman ranked correlations for the data shown in Figure 4.4. The numbers in parentheses indicate the plot shown in the figure, from top-left to bottom-right.*

Correlation Plot	Pearson correlation	Spearman ranked correlation
Linear (1)	0.998	0.999
Uniform (2)	−0.012	−0.011
Linear (3)	−0.999	−0.999
Circular (4)	0.000	0.003
Parabolic (5)	0.031	0.037
Sinusoid (6)	−0.766	−0.744

correlation coefficient obtained could be zero. One should be careful not to only rely on the computed value of ρ to determine if two variables are correlated or not. Not all non-linear patterns will have a zero correlation coefficient, as can be seen from the sinusoidal example shown, where $\rho = -0.77$.

As with the other variables mentioned, there is more than one type of correlation. If the type is not specified then it is assumed in this book that this corresponds to the Pearson correlation coefficient given by Eq. (4.23).

4.6.3 Spearman ranked correlation

An alternative correlation coefficient to the Pearson correlation discussed above is the Spearman ranked correlation. The computation method for the Spearman ranked correlation involves (i) ranking the values of variables on an entry-by-entry basis from least to greatest value, and then (ii) compute a correlation by combining the ranks of the individual variables. This way the absolute value is not used to compute the correlation coefficient, instead the rank of the data within the sample is used.

The Spearman ranked correlation coefficient is given by

$$\rho = 1 - \frac{6 \sum_{i=1}^{N} d_i^2}{N(N^2 - 1)}, \tag{4.28}$$

where d_i is the difference in rank of the values of x_i and y_i for the data, and N is the total number of events in the data set. The ranks take integer values from one to N. If tied ranks exists in the data, then a common value should be assigned. This

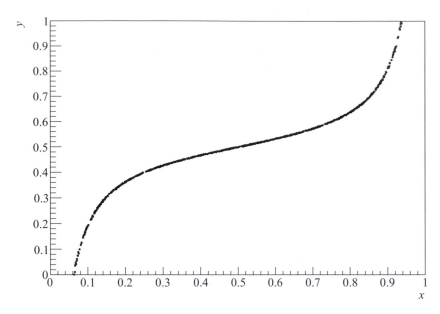

Figure 4.5 Data distributed according to $y = 0.5 + 0.1 \tan(\pi [x + 0.5])$. The Pearson correlation coefficient for these data is 0.941, whereas the Spearman ranked correlation is 1.000.

value is the average of ranks that would have been assigned if sequential numbers were given as ranks of the common valued data.

The Spearman ranked correlation coefficients computed for the data shown in Figure 4.4 are summarised (along with the Pearson correlation coefficients) in Table 4.1. One can see that for such distributions the computed correlation coefficients are very similar, and that both are insensitive to symmetry (for example the data distributed with a circular pattern). However, the usefulness of the Spearman ranked correlation comes into play when dealing with more complicated shapes. For example, consider the distribution shown in Figure 4.5. The Pearson correlation coefficient for these data is 0.941, whereas it is clear that the value of y for these data precisely follows changes in x. The corresponding Spearman ranked correlation is 1.000.

Example. Table 4.2 shows a set of data (x, y), along with the ranks (r_x, r_y), rank difference (d) and rank difference squared required to compute the Spearman ranked correlation. Two of the x values in the table have the same value, and instead of having to arbitrarily choose one to be rank 2 and the other to be rank 3, both are assigned a rank of 2.5 (as this is the average of ranks for the equal valued data). There are five data points, and the sum of squared rank differences is 6.5. Hence, the Spearman ranked correlation for these data as given by Eq. (4.28) is 0.675.

Table 4.2 *The data used to illustrate the computation of a Spearman ranked correlation as discussed in the text.*

x_i	y_i	r_{x_i}	r_{y_i}	d_i	d_i^2
0.1	0.1	1	1	0	0
0.5	0.2	4	2	2	4
0.2	0.3	2.5	3	−0.5	0.25
0.2	0.4	2.5	4	1.5	2.25
0.7	0.5	5	5	0	0

Numerically it is straightforward to compute the Spearman ranked correlation by hand for a few data. However, when dealing with large data samples, it is more convenient to use a computer programme with a suitable sorting algorithm. Several sorting algorithms with varying levels of efficiency are discussed by Press *et al.* (2002). More sophisticated sorting algorithms are reviewed in the third volume of Knuth (1998).

4.6.4 Removing correlations between variables

If pairs of variables are correlated, then manipulating the data in a consistent way can become complicated. Often it is possible to neglect the residual correlation between two variables x and y, while sometimes on neglecting correlations one introduces a significant bias in derived results. For this latter set of scenarios it is desirable to be able to remove the correlations between variables.

One way to approach this problem is to consider a two-dimensional rotation through some angle θ which can be described by the unitary transformation matrix

$$U = \begin{pmatrix} \cos\theta & \sin\theta \\ -\sin\theta & \cos\theta \end{pmatrix}. \tag{4.29}$$

A pair of coordinates (x, y) in the original space can be mapped onto a pair of coordinates in a transformed space (u, v) via a rotation through some angle θ described by

$$U\underline{x} = \begin{pmatrix} \cos\theta & \sin\theta \\ -\sin\theta & \cos\theta \end{pmatrix} \begin{pmatrix} x \\ y \end{pmatrix} = \begin{pmatrix} u \\ v \end{pmatrix}. \tag{4.30}$$

The inverse transformation can be obtained by applying $U^{-1} = U^T$ (as U us unitary) to the coordinate pair (u, v) to revert to the original basis and obtain the original coordinate (x, y). Given the form of U in Eq. (4.29) one can relate the covariance matrix V for the original dataset Ω_{xy} to the diagonal covariance matrix

V' of the transformed data set Ω_{uv} via

$$V = U^T V' U,$$

$$\begin{pmatrix} \sigma_x^2 & \sigma_{xy} \\ \sigma_{xy} & \sigma_y^2 \end{pmatrix} = U^T \cdot \begin{pmatrix} \sigma_u^2 & 0 \\ 0 & \sigma_v^2 \end{pmatrix} \cdot U. \tag{4.31}$$

Using the three unique simultaneous equations resulting from Eq. (4.31) it can be shown (see Exercise 4.13 at the end of this chapter) that

$$\theta = \frac{1}{2} \arctan \left(\frac{2\sigma_{xy}}{\sigma_x^2 - \sigma_y^2} \right). \tag{4.32}$$

Having determined θ for a given set of data, one can choose to transform the data set from a correlated (x, y) to an uncorrelated (u, v) basis. This has both advantages and disadvantages. One disadvantage is that the variables u and v will be a combination of x and y, and one may have to interpret any results obtained in terms of (u, v) back into (x, y). This may include the appropriate propagation of errors (see Chapter 6). A possible advantage of using such a transformation includes situations where one wishes to model the data in order to perform a fit (see Chapter 9). Here the lack of correlation between two variables will simplify the process of defining a model that is a good approximation of the data as one can define separate one-dimensional PDFs for u and v and compute their product $P(u)P(v)$ instead of having to devise a two-dimensional PDF $P(x, y)$ accounting for correlations, which is straightforward only in the case of a bivariate normal distribution (see Section 7.8.1). Another case where it may be advantageous to use uncorrelated variables is the process of averaging results (see Chapter 6).

Often one deals with data sets that have many variables, and so the two-dimensional rotation approach discussed above becomes tedious to apply and one needs to resort to using slightly more advanced linear algebra to remove correlations from sets of variables in the data. A more general solution to this problem can be readily seen by considering the covariance matrix between n variables $\underline{x}$. We wish to map the $\underline{x}$ space onto some un-correlated parameter space $\underline{u}$ in such a way that the $n \times n$ covariance matrix V' is diagonal. In order to do this we need to determine the transformation matrix required to diagonalise the covariance matrix V. Having done this we can apply the corresponding transformation on all events in the data set $\Omega(\underline{x})$ to produce the data set of rotated elements $\Omega'(\underline{u})$. Matrix diagonalisation is a well-known problem in linear algebra that one uses a number of methods of increasing rigour to perform. Given a matrix V, one can write down the eigenvalue equation

$$(V - \lambda I).\underline{r} = 0, \tag{4.33}$$

where λ is an eigenvalue, I is the identity matrix, and $\underline{r}$ is a vector of coordinates. The eigenvalues can be determined by solving $\det(V - \lambda I) = 0$, and given these it is possible to compute the corresponding eigenvectors, and hence V'. If this method fails to work, there are other methods including single value decomposition that may suffice, for example see Press *et al.* (2002).

To illustrate the use of this approach we can return to the two-dimensional case where one has a covariance matrix V given by

$$V = \begin{pmatrix} \sigma_x^2 & \sigma_{xy} \\ \sigma_{xy} & \sigma_y^2 \end{pmatrix}. \tag{4.34}$$

The eigenvalues of this matrix can be determined via

$$\det(V - \lambda I) = \begin{vmatrix} \sigma_x^2 - \lambda & \sigma_{xy} \\ \sigma_{xy} & \sigma_y^2 - \lambda \end{vmatrix}, \tag{4.35}$$

$$= \left(\sigma_x^2 - \lambda\right)\left(\sigma_y^2 - \lambda\right) - \sigma_{xy}^2, \tag{4.36}$$

$$= \lambda^2 - \left(\sigma_x^2 + \sigma_y^2\right)\lambda + \sigma_x^2\sigma_y^2 - \sigma_{xy}^2. \tag{4.37}$$

Thus the eigenvalues of the correlation matrix are given by

$$\lambda_\pm = \frac{\left(\sigma_x^2 + \sigma_y^2\right) \pm \sqrt{\left(\sigma_x^2 + \sigma_y^2\right)^2 - 4\left(\sigma_x^2\sigma_y^2 - \sigma_{xy}^2\right)}}{2}, \tag{4.38}$$

and one can determine the eigenvectors on noting that $V\underline{x}_\pm = \lambda_\pm\underline{x}_\pm$. Hence the (unnormalised) vectors $\underline{x}_\pm$ are given by

$$\underline{x}_+ = \begin{pmatrix} 1 \\ \left(\lambda_+ - \sigma_x^2\right)/\sigma_{xy} \end{pmatrix}, \text{ and } \underline{x}_- = \begin{pmatrix} 1 \\ \left(\lambda_- - \sigma_x^2\right)/\sigma_{xy} \end{pmatrix}. \tag{4.39}$$

If one combines the eigenvectors (ideally normalised to a magnitude of one for convenience) into a matrix $P = (\underline{x}_+, \underline{x}_-)$, then the diagonalised covariance matrix is

$$V' = P^{-1}VP = \begin{pmatrix} \lambda_+ & 0 \\ 0 & \lambda_- \end{pmatrix}. \tag{4.40}$$

Hence we can associate the eigenvalues of V with the variances of the variables $\underline{u}$ in the transformed space.

Table 4.3 *Data recorded from the experiment discussed in Section 4.7.1.*

x	0.10	0.22	0.25	0.50	0.55	0.70	0.80	0.90	1.00	1.11	1.12
y	0.9	1.1	1.2	1.2	1.3	1.4	1.5	1.3	1.6	1.5	1.3
z	0.1	−0.2	0.3	0.4	0.0	−0.4	0.1	−0.1	0.6	0.7	−0.3

4.7 Case study

This section discusses a case study related to the application of some of the techniques discussed in this chapter.

4.7.1 Analysing a data sample

Having performed an experiment, you measure three quantities x, y, and z for eleven data points which are reported in Table 4.3. From these you want to compute the mean, the error and the correlation matrices.

The arithmetic mean is given by Eq. (4.2), so as there are 11 data points, for each quantity you just add up the data and divide by 11 to obtain the following mean values: $\bar{x} = 0.659$, $\bar{y} = 1.30$, and $\bar{z} = 0.109$.

The error matrix is given by Eq. (4.21), so one must calculate both the variances (for the diagonal entries of the error matrix), and the covariance values (for the off-diagonal terms). There are only 11 data points, so before considering which variant of the equations one should use (with or without the Bessel correction factor), we can compare $1/n$ with $1/(n-1)$ to gauge the effect. For $n = 11$ the ratio of these two factors is 1.1, hence we should be using the variant with the Bessel correction factor to avoid a bias of 10% on the result. Hence the appropriate variance and covariance are given by Eqns (4.13) and (4.20), where one should replace n by $(n-1)$ for the latter. Using these results one obtains the following error matrix

$$V = \begin{pmatrix} 0.132 & 0.059 & 0.020 \\ 0.059 & 0.040 & 0.023 \\ 0.020 & 0.023 & 0.129 \end{pmatrix}. \tag{4.41}$$

The correlation matrix is related to the error matrix and the element ρ_{ij} is given by $V_{ij}/(\sigma_i \sigma_j)$, e.g. see Eq. (4.23). It follows from this that the diagonal elements of a correlation matrix will always be unity. The correlation matrix obtained from the data is given by

$$\rho = \begin{pmatrix} 1.000 & 0.816 & 0.154 \\ 0.816 & 1.000 & 0.320 \\ 0.154 & 0.320 & 1.000 \end{pmatrix}. \tag{4.42}$$

4.8 Summary

The main points introduced in this chapter are listed below.

(1) Data can be visualised in a number of ways, for example as a discrete histogram, a graph, or as some continuous function.

(2) In all cases, where appropriate one can visually represent uncertainties (or errors). The concept of an error will be discussed in detail in Chapter 6, however for the purposes of this chapter we consider the error to be given by the standard deviation, see Eq. (4.14).

(3) While the visual representation of data may be binned, one can describe data in terms of well-defined quantities computed on an event-by-event basis.

(4) A measurement is given by some notion of a nominal value, some indication of the spread of data and (when appropriate) units. For example, consider the following.

- Quantities that ascribe position information on an ensemble of data include the mean, median, and mode. The most useful of these quantities is the mean which is given by Eq. (4.2).
- The spread of data is usually represented by the variance (Eq. (4.13)) or standard deviation of the data as given by Eq. (4.14).
- A simple quantification of the width of a distribution is given by the full width at half maximum.
- The definitions of the spread or width of data discussed assume that the data are symmetrically distributed about some central value. Any asymmetry in the data can be quantified by computing the skew of the distribution.

(5) When working with data that are multi-dimensional, it is possible that some of the discriminating variables depend on each other. If this is the case, then we can describe the dependence in terms of the covariance or correlation between pairs of variables x and y. These quantities are discussed in Section 4.6.

(6) If any two variables are correlated, it is possible to remove that correlation by transforming the data (x, y) to some rotated space (x', y'), see Section 4.6.4. The ability to do this can be extremely useful for simplifying more advanced analysis of the data.

Exercises

4.1 Compute the mean μ, variance σ^2, standard deviation σ, and skew γ of the data sets Ω and κ (to 3 d.p.), where

$$\Omega = \{0.5, 0.9, 1.2, 1.5, 1.8, 2.0, 3.4, 4.1, 5.0, 5.1, 7.5, 8.5\},$$

and

$$\kappa = \{0.7, 0.8, 1.1, 1.2, 1.5, 1.8, 1.9, 2.0, 2.5, 2.6, 2.9, 3.5\}.$$

4.2 Compute the covariance matrix of the combined data set Ω and κ given in question 1, where Ω corresponds to x, and κ corresponds to y.

4.3 Given the covariance matrix in the previous question, compute the correlation matrix.

4.4 Compute the eigen values and eigen vectors corresponding to the error matrix V obtained in Exercise 4.2, thus determine the diagonalised form U of the error matrix.

4.5 Compute the mean, variance, standard deviation, and skew of the data set $\Omega(x) = \{1.0, 2.5, 3.0, 4.0, 4.5, 6.0\}$.

4.6 Compute the mean, variance, standard deviation, and skew of the data set $\Omega(x) = \{0.5, 1.0, 1.5, 1.6, 3.0, 2.1, 2.5\}$.

4.7 Compute the mean, standard deviation and correlation matrix for the following data.

x	0.10	0.22	0.25	0.50	0.55	0.70	0.80	0.90	1.00	1.11	1.12
y	1.0	1.1	1.1	1.2	1.3	1.4	1.4	1.3	1.6	1.5	1.4
z	0.1	−0.2	0.3	0.4	0.1	−0.4	0.1	−0.1	0.6	0.7	−0.3

4.8 Compute the mean, standard deviation and correlation matrix for the following data.

x	0.0	0.2	0.3	0.4	0.5	0.7	0.8	0.9	1.0	0.9	1.1
y	0.9	1.1	1.2	1.2	1.3	1.4	1.5	1.3	1.6	1.5	1.3
z	−0.1	−0.2	0.1	0.2	0.1	0.0	0.2	0.1	0.5	0.6	0.3

4.9 Compute the Spearman rank correlation coefficient for the following data.

x	0.5	0.7	0.8	0.9	1.1	1.3
y	0.9	0.8	1.1	1.2	1.2	1.0

4.10 Compute the Spearman rank correlation coefficient for the following data.

x	0.1	0.3	0.2	0.0	0.4	0.5	0.1	0.2	0.6	0.5
y	0.5	0.7	0.2	0.3	0.8	0.1	0.9	0.0	0.4	0.6

4.11 Starting from Eq. (4.31) determine expressions for both σ_u^2 and σ_v^2 hence θ.

4.12 Diagonalise the error matrix

$$V = \begin{pmatrix} 1.0 & 0.2 \\ 0.2 & 1.0 \end{pmatrix}. \tag{4.43}$$

4.13 Diagonalise the error matrix

$$V = \begin{pmatrix} 1.0 & 0.5 \\ 0.5 & 2.0 \end{pmatrix}. \tag{4.44}$$

4.14 Compute the rotation angle and matrix required to transform (x, y) to uncorrelated variables (u, v) given that $\sigma_x = 1.0$, $\sigma_y = 0.5$, and $\sigma_{xy} = 0.25$.

4.15 Compute the rotation angle and matrix required to transform (x, y) to uncorrelated variables (u, v) given that $\sigma_x = 2.0$, $\sigma_y = 1.5$, and $\sigma_{xy} = 1.0$.

5

Useful distributions

This chapter introduces four important distributions that can be used to describe a variety of situations. The first distribution encountered is that of the binomial distribution (Section 5.2). This is used to understand problems where the possible outcomes are binary, and usually categorised in terms of success and failure. For example, one can consider the situation of either detecting of failing to detect a particle passing through some apparatus as a binary event. The detection efficiency[1] in this particular problem is the parameter p of the binomial distribution. Typically one finds that $p \sim 1$ when working with efficient detectors. The Poisson distribution (Section 5.3) can be used to understand rare events where the total number of trials is not necessarily known, and the distribution depends on only the number of observed events and a single parameter λ that is both the mean and variance of the distribution. For example, the Poisson distribution can be used to describe the uncertainties on the content of each bin in Figure 1.2, which is a topic discussed in more detail in Chapter 7. The third distribution discussed here is the Gaussian distribution (Section 5.4). This plays a significant role in describing the uncertainties on measurements where the number of data are large. Finally the χ^2 distribution is introduced in Section 5.5. This distribution is typically encountered less frequently than the others, but still plays an important role in statistical data analysis. One of the uses of this distribution is to quantify the so-called goodness of fit between a model and a set of data. The probability determined for a given χ^2 and number of degrees of freedom can be a useful factor in determining if a fit result is valid, a topic encountered in Chapter 9.

These distributions are related to each other: in the limit of infinite data the binomial and χ^2 distributions tend to a Gaussian distribution. The Poisson distribution tends to a Gaussian distribution in the limit that $\lambda \to \infty$, and the binomial distribution is related to the Poisson distribution in the limit $p \to 0$. Additional

---

[1] The efficiency of detecting a particle is the ratio of the number of detected particles divided by the total number of particles. This topic is discussed in Section 6.3.

distributions are often encountered and a number of potentially useful ones are described in Appendix B. Section 5.1 introduces the formalism required to determine the expectation values of discrete and continuous distributions, which will be of use in the remainder of this chapter. It is worth noting that ones ability to correctly manipulate these distributions may vary depending on the algorithm used (see Section 5.6).

5.1 Expectation values of probability density functions

The notion of a probability density function (PDF) was introduced in Section 3.5. For some PDF denoted by $P(x)$ describing a continuous distribution, or $P(x_i)$ for a discrete distribution, we can compute the **expectation value** (the average value) of some quantity as the integral (or sum) over the quantity multiplied by the PDF. For example, the expectation value of the variable x, distributed according to the PDF $P(x)$ in the domain $-\infty < x < \infty$, is

$$\langle x \rangle = \int_{-\infty}^{+\infty} x P(x) dx, \tag{5.1}$$

in analogy with the discussion in Section 4.2. If we replace the variable x in Eq. (5.1) by a more complicated expression then we can compute the expectation values for other quantities. For example, the mean value of $V = (x - \overline{x})^2$ is given by

$$\langle V \rangle = \int_{-\infty}^{+\infty} (x - \overline{x})^2 P(x) dx. \tag{5.2}$$

The equivalent equations for a discrete distribution $P(x_i)$ are

$$\langle x \rangle = \sum_i x_i P(x_i), \tag{5.3}$$

and

$$\langle V \rangle = \sum_i (x_i - \overline{x})^2 P(x_i), \tag{5.4}$$

where the sum is over all bins. These results will be useful throughout the remainder of this chapter.

5.2 Binomial distribution

Consider the case of flipping an unbiased coin as described in Section 3.7.1. There are two possible outcomes; H and T. If we choose H, and flip a coin, the probability of a success p is 0.5, and the probability of a failure $(1 - p) = q$ is also 0.5. We can try flipping the coin a number of times n. For each coin flip there are two possible outcomes: success and failure, thus there are 2^n possible permutations of flipping the coin. It is possible to compute the number of combinations of obtaining r successes from n trials as

$$^nC_r = \frac{n!}{r!(n-r)!}. \tag{5.5}$$

The probability of success and failure are equal for flipping an unbiased coin, and we previously assigned them values of p and $1 - p$, respectively. We can multiply the number of possible permutations as given by ^nC_r by the probability of r successes and $n - r$ failures in order to obtain

$$P(r; p, n) = p^r(1 - p)^{n-r} \frac{n!}{r!(n-r)!}. \tag{5.6}$$

The result $P(r; p, n)$ is the probability of obtaining r successes from n experiments where the probability of a successful outcome of an experiment is given by p. We have seen this result before in the context of flipping a coin (see Chapter 3). Now we have obtained a generalised solution.

The ***binomial distribution*** is the distribution corresponding to the probabilities computed using Eq. (5.6) and is a function of r, p, and n. In general this can be used to describe the probability of the number of possible outcomes when repeating an experiment with binary output a number of times. Figure 5.1 shows the binomial distribution expected for several different cases of n and p. Note that when $p = 0.5$, and $q = 0.5$, the distribution obtained is symmetric about the mean value. The mean and variance of a binomial probability distribution are given by (see Section 5.2.1 for the proof of these results)

$$\langle r \rangle = np, \tag{5.7}$$
$$V(r) = np(1 - p). \tag{5.8}$$

Hence the standard deviation of a binomial distribution is $\sigma = \sqrt{np(1 - p)}$.

Example. Consider two coins – one being unbiased with $p = 0.5$, and the other being biased toward heads with $p = 0.7$. If we flip both coins ten times, what is the probability that we obtain equal numbers of heads (H) and tails (T) for each coin?

Firstly, let's consider how many combinations of H and T there are in order to obtain this result. This is given by Eq. (5.5) where $n = 10$ and $r = 5$, so the number of combinations we are interested in is

$$^{10}C_5 = \frac{10!}{5!5!}. \tag{5.9}$$

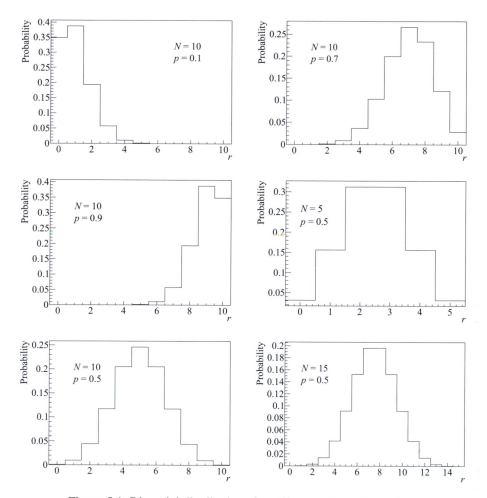

Figure 5.1 Binomial distributions for different values of n and p.

This can be used to compute the probability of obtaining $5H5T$ with a fair coin, using Eq. (5.6) which is

$$P(5; 0.5, 10) = 0.5^5(1 - 0.5)^5 \frac{10!}{5!5!}. \tag{5.10}$$

$$= 0.246. \tag{5.11}$$

If we use the second coin, then as $p = 0.7$ we obtain a probability of $P(5; 0.7, 10) = 0.103$. There is a factor of 2.4 difference in the probability of obtaining equal numbers of heads compared to tails when using the biased coin instead of the fair coin. The mean number of heads obtained with the fair coin is five; however, if we use the biased coin we would find that the mean number of heads obtained would be seven. The distributions of possible outcomes as a function of r (the number of heads obtained) for these experiments are shown in Figure 5.1. The top-right figure is for the biased coin, and the bottom-left one is for the unbiased coin. Tables of binomial probabilities can be found in Appendix E.1.

Another way of viewing the binomial distribution can be illustrated in terms of the concept of a testing for a biased coin. Instead of considering absolute probabilities one can study the likelihood distribution and compare possible outcomes. If we perform an experiment where we flip a coin that we suspect is biased some number of times n, then we can use the number of observed heads r to compute the probability $P(p; n, r) \propto p^r (1 - p)^{n-r}$. The likelihood distribution $L(p; n, r)$ is obtained when the maximum value of $P(p; n, r)$ is set to unity. Figure 5.2 shows the likelihood of p obtained for r heads for five and 15 trials. If you flipped a coin five times and got zero heads, you might start to become suspicious that the coin could be biased, but you would not be able to rule out the possibility that the coin was really unbiased. One can, however, compare the likelihood of an unbiased coin against one with a given bias, and compute a likelihood ratio of the two possible outcomes. If on the other hand you flip a coin 15 times and only obtain tails, then you would be left with two possibilities. Either the coin is biased, or your 15 trials led to an extremely unlucky outcome. The former conclusion is the most likely one given the available data; however, it should be noted that while in this case the likelihood that the coin is unbiased is very small, it is not zero. Hence it is possible to conclude that the coin is biased only if one is prepared to accept the possibility that there is a chance that the conclusion may be incorrect. This issue exemplifies the need for hypothesis testing, which is discussed in Chapter 8.

5.2.1 Derivation of the mean and variance of the binomial distribution

The binomial distribution is given by Eq. (5.6). For this to be valid, the total probability of anything to happen should be unity. In other words, it should be certain that something happens for a given value of p, and a given number of trials n. So we can sum Eq. (5.6) over r to verify this property

$$\sum_{r=0}^{r=n} P(r; p, n) = \sum_{r=0}^{r=n} p^r (1 - p)^{n-r} \frac{n!}{r!(n-r)!}, \tag{5.12}$$

$$= (1 - p)^n + np(1 - p)^{n-1} + \frac{n(n-1)}{2!} p^2 (1 - p)^2 + \cdots$$

$$+ np^{n-1}(1 - p) + p^n. \tag{5.13}$$

One can prove by induction (for $n = 0, 1, 2, 3, \ldots$, thus for arbitrary n) that the sum of this binomial expansion is always unity.

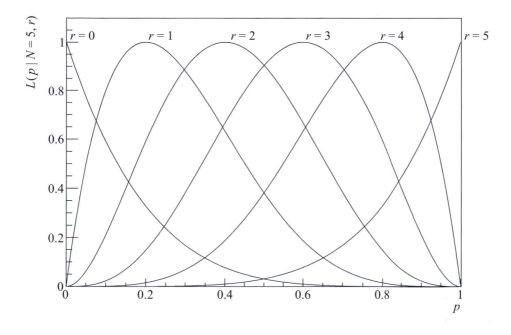

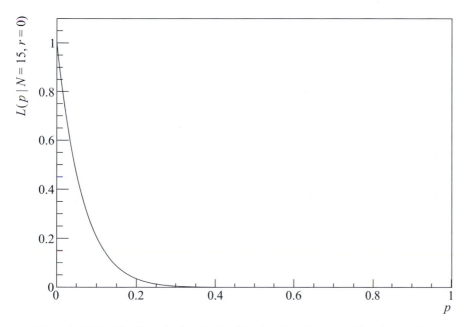

Figure 5.2 The likelihood of p obtained for (top) a coin tossed five times, obtaining r heads, and (bottom) a coin tossed 15 times, obtaining no heads.

Following on from Eq. (5.3), the mean value of r is given by

$$\langle r \rangle = \sum_{r=0}^{r=n} r P(r; p, n), \tag{5.14}$$

$$= \sum_{r=0}^{r=n} r p^r (1-p)^{n-r} \frac{n!}{r!(n-r)!}, \tag{5.15}$$

$$= np \sum_{r=0}^{r=n} p^{r-1} (1-p)^{n-r} \frac{(n-1)!}{(r-1)!(n-r)!}, \tag{5.16}$$

$$= np \sum_{r=0}^{r=n} P(r; p-1, n-1), \tag{5.17}$$

$$= np, \tag{5.18}$$

as the sum over all possible outcomes of the binomial distribution is unity. In order to compute the variance on r, one needs to use the above result. The starting point to determine $V(r)$ is

$$V(r) = \sum_{r=0}^{r=n} (r - np)^2 P(r; p, n), \tag{5.19}$$

$$= \langle r^2 \rangle - \langle r \rangle^2, \tag{5.20}$$

$$= \langle r^2 \rangle - n^2 p^2. \tag{5.21}$$

In order to compute $\langle r^2 \rangle$, in analogy with the derivation above for $\langle r \rangle$, one needs to absorb the factor of r^2 into the $r!$ term in the binomial series. It is not possible to do this directly, however one can absorb a factor of $r(r-1)$ into that sum, and hence compute $V(r)$ via

$$V(r) = \langle r(r-1) \rangle + \langle r \rangle - \langle r \rangle^2. \tag{5.22}$$

As anticipated, the right-hand side of the previous equation reduces to the desired result of $np(1-p)$.

5.3 Poisson distribution

The ***Poisson distribution*** is given by

$$P(r, \lambda) = \frac{\lambda^r e^{-\lambda}}{r!}, \tag{5.23}$$

which is a function of the number of observed events r, and λ. The parameter λ is the mean and variance of the distribution (as shown in Section 5.3.1). Figure 5.3

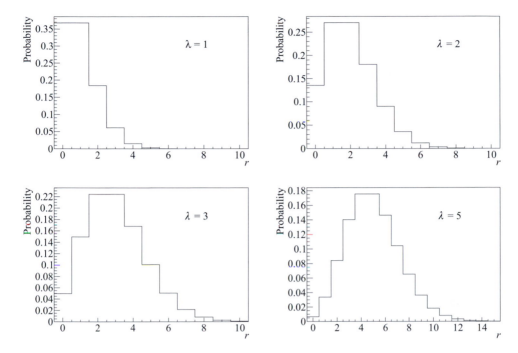

Figure 5.3 Poisson probability distributions for different values of λ.

shows the Poisson probability distribution for several different values of λ. For small λ the distribution is asymmetric and skewed to the right. As λ increases the Poisson distribution becomes more symmetric.

This is an important distribution often used in science when studying the occurrence of rare events in a continually running experiment. For example, the Poisson distribution can be used to describe radioactive decay (see Section 1.3) or particle interactions where we have no idea how many decays occur in total, but can study events over a finite time to understand the underlying behaviour. Tables of Poisson probabilities can be found in Appendix E.2.

If one considers the situation where r events have been observed, one can ask the following question; what is the most likely value of λ corresponding to this observation? Figure 5.4 shows the corresponding likelihood distribution for $r = 0, 1, 2, 3, 4$, and 5 events. One can see that for $r > 0$ the Poisson distribution has a definite non-zero maximum to the likelihood distribution.

5.3.1 Derivation of the mean and variance of the Poisson probability distribution

If one considers the Poisson probability distribution as given by Eq. (5.23), then the sum of this distribution over all values of r is unity. This can be shown as

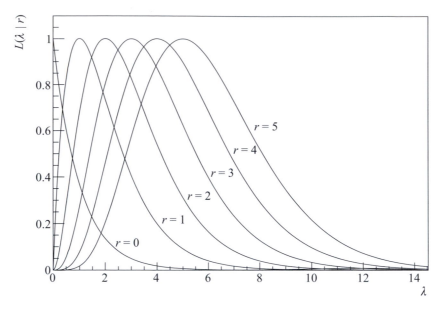

Figure 5.4 Likelihood distributions corresponding to Poisson probability distributions for different values of r as a function of λ.

follows

$$P(r, \lambda) = \frac{\lambda^r e^{-\lambda}}{r!}, \tag{5.24}$$

$$\sum_{r=0}^{r=\infty} P(r, \lambda) = e^{-\lambda} \left[1 + \lambda + \frac{\lambda^2}{2!} + \frac{\lambda^3}{3!} + \frac{\lambda^4}{4!} + \cdots \right], \tag{5.25}$$

$$= e^{-\lambda} e^{\lambda}, \tag{5.26}$$

$$= 1, \tag{5.27}$$

as the term in the square brackets is the Maclaurin series expansion for e^{λ}. As required the Poisson probability distribution is normalised to unity.

Following on from Section 5.1, the mean value of r is given by

$$\langle r \rangle = \sum_{r=0}^{r=\infty} r P(r, \lambda), \tag{5.28}$$

$$= \sum_{r=0}^{r=\infty} r \frac{\lambda^r e^{-\lambda}}{r!}, \tag{5.29}$$

$$= 0 + \lambda e^{-\lambda} + \frac{2\lambda^2 e^{-\lambda}}{2!} + \frac{3\lambda^3 e^{-\lambda}}{3!}, \tag{5.30}$$

$$= \lambda e^{-\lambda} \left[1 + \lambda + \frac{\lambda^2}{2!} + \cdots \right], \tag{5.31}$$

$$= \lambda. \tag{5.32}$$

Similarly the variance of r is given by

$$V(r) = \sum_{r=0}^{r=\infty} (r - \lambda)^2 P(r, \lambda), \tag{5.33}$$

$$= \sum_{r=0}^{r=\infty} (r - \lambda)^2 \frac{\lambda^r e^{-\lambda}}{r!}, \tag{5.34}$$

$$= \lambda e^{-\lambda} \left[\lambda + (1 - \lambda)^2 + \frac{(2 - \lambda)^2 \lambda}{2!} + \frac{(3 - \lambda)^2 \lambda^2}{3!} + \cdots \right]. \tag{5.35}$$

After gathering together the terms in the square bracket, these simplify to the Maclaurin series expansion for e^λ and we find

$$V(r) = \lambda e^{-\lambda} e^\lambda, \tag{5.36}$$

$$= \lambda. \tag{5.37}$$

Thus we obtain the expected results that the PDF is normalised to a total probability of one, and that the distribution has a common mean and variance given by λ.

5.4 Gaussian distribution

The **Gaussian distribution**, also known as the **normal distribution**, with a mean value μ and standard deviation σ as a function of some variable x is given by

$$P(x, \mu, \sigma) = \frac{1}{\sigma \sqrt{2\pi}} e^{-(x-\mu)^2/2\sigma^2}. \tag{5.38}$$

This distribution will be discussed in more detail in Chapter 6 when considering the nature of statistical errors. Figure 5.5 shows a Gaussian distribution with $\mu = 0$ and $\sigma = 1$. Often it is useful to transform data from the x space to a corresponding z space which has a mean value of zero, and a standard deviation of one. This transformation is a one to one mapping given by

$$z = \frac{x - \mu}{\sigma}, \tag{5.39}$$

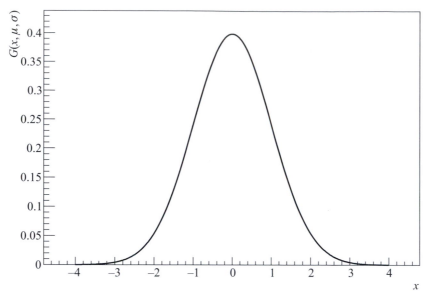

Figure 5.5 Gaussian distribution with a mean value $\mu = 0$ and width $\sigma = 1$. This is equivalent to $P(z)$ given in Eq. (5.40).

for any given Gaussian distribution $G(x, \mu, \sigma)$. Hence the Gaussian distribution in terms of z is

$$P(z) = \frac{1}{\sqrt{2\pi}} e^{-z^2/2}. \tag{5.40}$$

There are a number of useful results associated with integrals of the Gaussian distribution. These are discussed in the following, and can be used in order to determine the confidence intervals discussed in Chapter 7. Firstly the normalisation condition for a Gaussian distribution is given by

$$I = \frac{1}{\sigma\sqrt{2\pi}} \int\limits_{-\infty}^{+\infty} e^{-(x-\mu)^2/2\sigma^2} dx, \tag{5.41}$$

$$= 1. \tag{5.42}$$

While it is possible to analytically integrate the Gaussian distribution, it is often convenient to perform a numerical integration instead (see Appendix C for a short introduction to this subject). Tables E.8 and E.9 summarise the integral of a Gaussian function from $-\infty$ to a given number of standard deviations above the mean, and within a given number of standard deviations above and below the mean, respectively.

One can extend the Gaussian distribution to encompass more than one dimension, resulting in the so-called multi-normal distribution, or multivariate Gaussian distribution. This distribution is discussed in Section 7.8.1 in the context of confidence intervals corresponding to an *error ellipse* resulting from two correlated parameters.

5.5 χ^2 distribution

The χ^2 *distribution* is given by

$$P(\chi^2, v) = \frac{2^{-v/2}}{\Gamma(v/2)}(\chi^2)^{(v/2-1)}e^{-\chi^2/2}, \tag{5.43}$$

as a function of χ^2, and the number of **degrees of freedom** v. The χ^2 sum is over squared differences between data x_i and a hypothesis $\widehat{x}$, normalised by the uncertainty on the data $\sigma(x_i)$. Hence

$$\chi^2 = \sum_{i=data} \left[\frac{x_i - \widehat{x}}{\sigma(x_i)}\right]^2. \tag{5.44}$$

The quantity v is given by the number of entries in the data sample less any constraints imposed or adjusted to compute the χ^2. The quantity $\Gamma(v/2)$ is the **gamma distribution** which has the form

$$\Gamma(v/2) = (v/2 - 1)!, \tag{5.45}$$

and is valid for positive integer values of v. The gamma function for $v = 1$ is $\Gamma(1/2) = \sqrt{\pi}$.

For example, if we have a sample of n data points, then there are $n - 1$ degrees of freedom (as the total number of data points is itself a constraint), so $v = n - 1$. Given some model, we can then compute the value of χ^2 using Eq. (5.44), and thus $P(\chi^2, v)$ using Eq. (5.43). Figure 5.6 shows examples of the χ^2 distribution for different numbers of degrees of freedom. The χ^2 distribution can be used to validate the agreement between some model and a set of data. As a general rule of thumb, a reasonable agreement between a set of data $\Omega(x)$ and some model of the data $f(x)$ is obtained for $\chi^2/v \sim 1$. The χ^2 probability distribution enables one to go beyond applying a simple rule of thumb and allows the experimenter to compute a probability associated with the agreement between data and model. More generally this type of comparison is often referred to as **goodness of fit** (GOF) test.

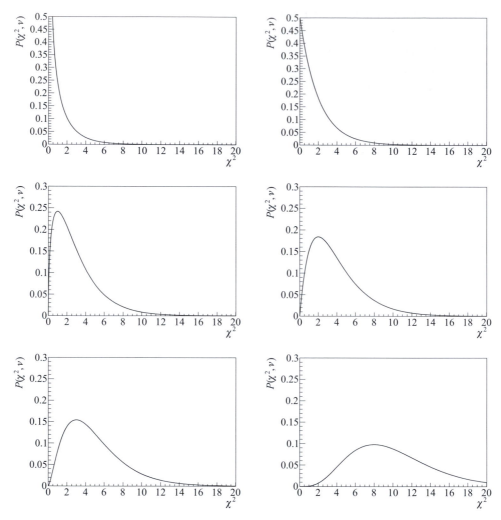

Figure 5.6 Examples of the χ^2 probability distribution for $\nu = 1, 2, 3, 4, 5$, and 10 (left to right, top to bottom). Note that the vertical scales are not the same for all plots.

5.6 Computational issues

When a distribution involves the computation of very large or very small numbers, then it is usually necessary to take care in the calculation. A classic example of this is the computation of n factorial ($n!$). If we use a computer to calculate n! for us with a 4-bit integer representation, then we will find that we can calculate $13! = 1\,932\,053\,504$, but we obtain a solution for 14! that is smaller than 13!, which is clearly not correct. The reason that we compute the wrong answer is that we run out of memory allocated for the representation of the integer result. We can use a longer integer representation for this calculation, however, we will run into the same problem once again for some larger value of n – so we may only temporarily postpone the problem. When a situation like this is encountered, a computer will not give us a warning, and will merrily continue to do what it is told, so it is up to the person doing the calculation to make sure that the result is sensible.

A second illustration of computational problems that can be encountered is given by the binomial distribution

$$P(r; p, n) = p^r(1 - p)^{n-r}\frac{n!}{r!(n - r)!}. \tag{5.46}$$

In addition to having to compute factorials that we know from above can be problematic, we also have to compute powers of numbers less than one. As either r or $n - r$ become large, then the powers of p and q we need to compute may quickly become small, and here we run into another problem. If we represent a real number with some floating point form in a computer, then that too has a finite resolution. We may be able to compute p^r and q^{n-r}, but find that the product of the two small but finite numbers is zero. In such instances, it can be beneficial to multiply the small numbers by a large factor to compute the probability, then divide out those large factors at the end of the calculation so that the total probability for anything to happen remains normalised to unity. Again it is up to the person doing the calculation to ensure that the result obtained is correct, and any scale factors introduced into a computation are sufficient to compute the result to the desired level of accuracy.

For such cases it would be desirable if we could use a more robust algorithm, rather than postpone problems resulting from the precision of number representation in a computer by buying a new computer or compiler with a larger bit representation for floating point or integer numbers. Indeed often it is possible to find a more robust algorithm to compute a quantity with a little thought: take, for example, the case of the Poisson probability distribution

$$P(r, \lambda) = \frac{\lambda^r e^{-\lambda}}{r!}. \tag{5.47}$$

If we compute $P(0, \lambda)$, then we can obtain $P(1, \lambda)$ by multiplying our result for $P(0, \lambda)$ by λ and dividing by the incremented value of r (in this case by one). So in general we can use the following rule to compute successive terms of a Poisson probability distribution

$$P(r + 1, \lambda) = \frac{\lambda P(r, \lambda)}{r + 1}. \tag{5.48}$$

By using this algorithm we never have to compute the factorial of a large number, and at each step in the calculation we find that we obtain sensible results, even for large r. Common numerical issues such as the ones presented here are discussed in literature on computer algorithms, e.g. see volume one of Knuth (1998).

5.7 Summary

The main points introduced in this chapter are listed below.

(1) Probability density functions are versatile tools that can be used to describe common scenarios. There are several very important PDFs described in this chapter; these are distributions which are intimately linked with the concepts of values and uncertainties of measurements and thus play an important role in understanding what information can be obtained from data.

  - The binomial distribution can be used to describe the outcome of an event if the probability p of success is known, and the number of trials n is also determined. The probability of a given outcome r for such a scenario is given by Eq. (5.6).
  - The Poisson distribution can be used to describe the probability of a rare event occurring with some mean and variance as given by λ, for some unknown number of trials. The probability for such an event to occur is given by Eq. (5.23).
  - The Gaussian distribution is intimately related to the definition of uncertainty, and the functional form of this distribution is given in Eq. (5.38). The relationship between this function and experimental uncertainty is discussed in Chapter 6.
  - The χ^2 distribution is commonly used in data analysis to compare the data to a model, and understand if the optimal solution obtained is meaningful or not. The use of the χ^2 in fitting is discussed at length in Chapter 9, and the χ^2 probability distribution is given by Eq. (5.43).
  - Note that additional PDFs are given in Section 7.8.1 and Appendix B.

(2) A significant practical issue was raised: that of computing and manipulating quantities. When computing probabilities, or related quantities, it is vital that the person doing the calculation understands the limitations of the calculating device and/or algorithm used. Familiarity with computational tools will help one identify the potential limitations affecting a calculation. It can be easy to overlook the subtleties of implementing an algorithm in the form of a computer program, and care should be taken when doing this in order to ensure that the results obtained are sensible.

Exercises

5.1   Determine the binomial probability associated with flipping an unbiased coin five times and obtaining either no heads or all heads.

5.2 Determine the binomial probability associated with flipping a biased coin with $p = 0.4$ five times and obtaining either no heads or all heads.

5.3 Determine the binomial probability associated with flipping a biased coin with $p = 0.4$ five times and obtaining three heads and two tails.

5.4 Determine the Poisson probability to observe zero signal events in an experiment given $\lambda = 3$.

5.5 Determine the cumulative Poisson probability to observe at least three signal events in an experiment given $\lambda = 3$.

5.6 Compute the ratio of the likelihood of observing one event with respect to that of observing five events given a Poisson distribution with $\lambda = 4$.

5.7 Determine the probability contained within $\pm 1, \pm 2$, and $\pm 3\,\sigma$ for a Gaussian distribution.

5.8 What z value has approximately 90% of the probability below it, and 10% of the probability above it, for a Gaussian distribution?

5.9 What is the likelihood for $z = \pm 1\sigma$ for a Gaussian distribution, assuming $\mathcal{L}_{max} = 1$?

5.10 What is the probability for $\chi^2 = 5$ given four degrees of freedom? Is the result reasonable?

5.11 What is the probability for $\chi^2 = 1$ given ten degrees of freedom? Is the result reasonable?

5.12 What is the probability for $\chi^2 = 8$ given two degrees of freedom? Is the result reasonable?

5.13 Compute the mean and variance of the sum of two Poisson distributions with means of λ_1 and λ_2, respectively. What do you conclude from this?

5.14 Compute the expectation value of some variable x distributed according to a probability density function $P(x) = a + e^{-x}$, over the domain $0 \le x \le 1$.

5.15 Compute the variance of x distributed according to the probability density function given in the previous exercise.

6

Uncertainty and errors

6.1 The nature of errors

We now have enough background information to revisit the topic of quantifying the uncertainty on a single, or set, of measurement(s). The concepts developed in this chapter are central to how scientific method is able to test theory against experimental observations. In practice, the words error and uncertainty are often used interchangeably when describing how well we are able to constrain or determine the value of an observable.

6.1.1 Central limit theorem

If one takes N random samples of a distribution of data that describes some variable x, where each sample is independent and has a mean value μ_i and variance σ_i^2, then the sum of the samples will have a mean value M and variance V where

$$M = \sum_{i=1}^{N} \mu_i, \qquad (6.1)$$

$$V = \sum_{i=1}^{N} \sigma_i^2. \qquad (6.2)$$

As N tends to infinity, the distribution of the sum of the samples tends to a Gaussian distribution. This is called the ***central limit theorem*** (CLT) and it implies that independent measurements of some observable will have uncertainties that are Gaussian in nature. Indeed empirical tests of this theorem show that this is indeed the case. As shown, for example, in Barlow (1989), Eqs. (6.1) and (6.2) can be verified by expanding the right-hand sides of those equations and substituting in Eqs. (4.2) and (4.5), respectively.

Table 6.1 *The Gaussian probability for some repeat measurement of an observable x to lie between +nσ and −nσ of the original measurement for values of n up to n = 5, shown to a precision of four decimal places.*

$\pm n\widehat{\sigma}$	coverage (%)
1	68.2689
2	95.4500
3	99.7300
4	99.9937
5	99.9999

6.1.2 The Gaussian nature of statistical uncertainty

Statistical uncertainties are Gaussian in nature for sufficiently large samples of data. This result can be obtained both empirically and via the CLT. The ramification of this result will be discussed in more detail in the rest of this section; however, before looking at this in detail, we briefly return to our expression of a measurement $\widehat{x} \pm \widehat{\sigma}$, where $\widehat{x}$ is our best estimate of the true mean value, and $\widehat{\sigma}$ is the estimate of the uncertainty on $\widehat{x}$ which is given by the standard deviation of the result. We interpret this result as saying that if we were to repeat the measurement, there is a certain probability that the second measurement would be within $\pm 1\sigma$ of the original result. This probability is given by

$$P = \int_{-1\widehat{\sigma}}^{+1\widehat{\sigma}} G(x;\widehat{x},\widehat{\sigma})dx, \qquad (6.3)$$

where $G(x;\widehat{x},\widehat{\sigma})$ is a Gaussian PDF with mean $\widehat{x}$ and width $\widehat{\sigma}$. This probability is approximately 68.3%, and is sometimes referred to as the ***coverage*** of the interval $(x - \widehat{x}) \in [-1\sigma, +1\sigma]$. Similarly we can consider how often we would expect the second result to be within $\pm n\widehat{\sigma}$ of the original experiment. These results are summarised in Table 6.1 for $n = 1, 2, \ldots 5$.

If we perform a measurement on some observable O, the result of which is given by $O_1 = \widehat{x}_1 \pm \widehat{\sigma}_1$, then at some later time we repeat the measurement with the same experimental conditions, we would expect that, by definition, we can find the new measurement $O_2 = \widehat{x}_2 \pm \widehat{\sigma}_2$, which can have a central value $\widehat{x}_2$ that is more than σ_1 away from $\widehat{x}_1$. If we perform an infinite number of subsequent measurements of the observable O we would expect that 68.2689% of the time, the results will fall between $\widehat{x}_1 - \widehat{\sigma}_1$ and $\widehat{x}_1 + \widehat{\sigma}_1$, and thus 31.7311% of the time the results will fall outside of this range. Implicitly when doing this we assume that $\widehat{x}_1$ is a sufficiently good estimate of the true mean value. By construction we are expecting that when we repeat a measurement, quite a lot of the time our new result will be slightly different than our original one. Normally we consider two results to be compatible if a subsequent measurement is within $\pm 3\sigma$ of the original on the basis that we usually make many measurements. Therefore it is not unreasonable to expect that less than one in a hundred results has a large ($\geq \pm 3\sigma$) deviation from our expectations. Much more than this level of discrepancy and we would want to investigate the compatibility of the two results more rigorously in order to verify that any differences can be attributed to statistical variations in the measurements.

6.1.3 Repeating measurements with a more precise experiment

Often in science, we attempt to repeat the measurement of an observable in order to obtain a more stringent constraint on that observable than obtained by any previous experiment. An example of this would be to measure a physical observable, such as the mass of an electron, with greater precision. When doing so, in addition to preparing or constructing a more robust experiment, it may be necessary to increase the statistical power of the experiment. By 'increase the statistical power' we mean, increase the number of data entries recorded by the repeat experiment.

So the following question arises; what would constitute a minimum increase in the number of data entries or events recorded at an experiment in order to justify the time and expense of building a new one? To answer this question, you have to understand how much better your new experiment would be relative to the old one(s) and offset the cost and time required to construct a new experiment and repeat the measurement against the benefit obtained in terms of increased precision. The increase in precision expected can be determined from the definition of standard deviation given in Eq. (4.14). The statistical precision of a new experiment with 10 times the data of an old one would have an error $\sqrt{10} \sim 3.2$ times smaller than the old experiment. So if the statistical uncertainty dominated the precision of the measurement you wanted to make, then it would be worth considering making the new experiment. Much less than this factor of increase in precision, and one could spend a lot of time working on a new measurement that provides only a very marginal improvement over the previous result. If one were to build an experiment that could collect 100 times the data of any existing experiment, then that new endeavour would be able to make a measurement 10 times better than any existing one. In general we follow the rule that statistical precision will scale with the square-root of the number of data, or $\sqrt{N}$.

If a great deal of care is taken in the design and construction of the new experiment, then it may be possible to obtain a better than $\sqrt{N}$ reduction in the error.
But that would require the new experiment to be significantly better than the old
one (for example, better energy or momentum resolution, better calibration, etc.).
If there is no such improvement and someone claims to be able to do better than
the $\sqrt{N}$ scaling in uncertainty by collecting more data with an equivalent method
to that used in an existing experiment, then one should seriously question how this
new measurement can 'beat statistics'. It is not possible to 'beat statistics', hence
this scaling rule can be used to make, and check the validity of, extrapolations from
an old experiment to a new one.

6.2 Combination of errors

In order to understand how to combine uncertainties we can first consider the case
of a function dependent on two measured inputs: x and y each with with mean
values $\bar{x}$ and $\bar{y}$, and errors σ_x and σ_y. We can Taylor expand this function $f(x, y)$
about $x = \bar{x}$, and $x = \bar{y}$ as

$$f(x, y) = f(\bar{x}, \bar{y}) + \frac{\partial f}{\partial x}(x - \bar{x}) + \frac{\partial f}{\partial y}(y - \bar{y}) + \cdots , \tag{6.4}$$

where in the following we ignore higher-order terms. If we now consider how to
compute the variance of some quantity f as described in Section 4.3, this is given
by

$$V(f) = (f - \bar{f})^2, \tag{6.5}$$

where $\bar{f} = f(\bar{x}, \bar{y})$. Thus, it follows that

$$V(f) = \left(\frac{\partial f}{\partial x}(x - \bar{x}) + \frac{\partial f}{\partial y}(y - \bar{y}) \right)^2 , \tag{6.6}$$

$$= \left(\frac{\partial f}{\partial x} \right)^2 \sigma_x^2 + \left(\frac{\partial f}{\partial y} \right)^2 \sigma_y^2 + 2 \frac{\partial f}{\partial x} \frac{\partial f}{\partial y} \sigma_{xy}, \tag{6.7}$$

where we have replaced $(x - \bar{x})^2$, $(y - \bar{y})^2$, and $(x - \bar{x})(y - \bar{y})$ by σ_x^2, σ_y^2, the
covariance σ_{xy}, and neglected higher order terms in the last step. Equation (6.7)
can be expressed in matrix form as

$$V(f) = \left(\frac{\partial f}{\partial x}, \frac{\partial f}{\partial y} \right) \begin{pmatrix} \sigma_x^2 & \sigma_{xy} \\ \sigma_{xy} & \sigma_y^2 \end{pmatrix} \begin{pmatrix} \frac{\partial f}{\partial x} \\ \frac{\partial f}{\partial y} \end{pmatrix} , \tag{6.8}$$

where the covariance matrix V (Eq. 4.22) is evident as the middle term in the
equation.

If x and y are independent variables, the third term in Eq. (6.7) vanishes, and we obtain a general expression for combining uncertainties for functions of two independent variables

$$\sigma_f^2 = \left(\frac{\partial f}{\partial x}\right)^2 \sigma_x^2 + \left(\frac{\partial f}{\partial y}\right)^2 \sigma_y^2. \tag{6.9}$$

The previous discussion can be generalised in order to compute the uncertainty on some function with an arbitrary number of input variables $f = f(x, y, z, \ldots)$ where the observables $u_i = x, y, z, \ldots$ are independent, resulting in

$$\sigma_f^2 = \left(\frac{\partial f}{\partial x}\right)^2 \sigma_x^2 + \left(\frac{\partial f}{\partial y}\right)^2 \sigma_y^2 + \left(\frac{\partial f}{\partial z}\right)^2 \sigma_z^2 + \cdots, \tag{6.10}$$

$$= \sum_i \left(\frac{\partial f}{\partial u_i}\right)^2 \sigma_{u_i}^2, \tag{6.11}$$

and the sum is over the observables.

So given the variances and values of the variables u_i, as well as the form of the differentiable functional f it is possible to determine the variance of f. This general form can be used in order to determine some familiar results, as shown in the following. We shall return to the scenario of correlated observables in Section 6.2.4.

6.2.1 Functions of one variable

If the function f has a form that depends only on one observable x, for example

$$f = ax + b \tag{6.12}$$

we can use Eq. (6.11), where here the sum is over a single term ($i = 1$) and $u_1 = x$, to determine how the error on x is related to the error on f. This is given by

$$\sigma_f = \sqrt{\left(\frac{\partial f}{\partial x}\right)^2 \sigma_x^2} \tag{6.13}$$

$$= a\sigma_x. \tag{6.14}$$

So for this function, the error on f, given by σ_f, is a times the error on x. This result is independent of any offset of the measured observable. This can be cross-checked by starting from the definition of variance given in Eq. (4.5) and replacing x with

$ax + b$ as follows

$$V(x) = \frac{1}{n} \sum_{i=1}^{n} (ax_i + b - \overline{ax + b})^2, \tag{6.15}$$

$$= \langle (ax + b)^2 \rangle - \langle ax_i + b \rangle^2, \tag{6.16}$$

$$= \langle a^2 x^2 + 2abx + b^2 \rangle - a^2 \langle x \rangle^2 - 2ab\langle x \rangle - b^2, \tag{6.17}$$

$$= a^2 \langle x^2 \rangle + 2ab\langle x \rangle + b^2 - a^2 \langle x \rangle^2 - 2ab\langle x \rangle - b^2, \tag{6.18}$$

$$= a^2 \langle x^2 \rangle - a^2 \langle x \rangle^2, \tag{6.19}$$

$$\sqrt{V(x)} = a\sigma_x = \sigma_f. \tag{6.20}$$

In both cases we obtain the same result, as required.

6.2.2 Functions of two variables

Now consider the function $f = x + y$, where we have measured the mean and standard deviation of both x and y, and want to compute the standard deviation on their sum. We can use the general formula of Eq. (6.11) to determine how to do this, hence

$$\sigma_f^2 = \left(\frac{\partial f}{\partial x}\right)^2 \sigma_x^2 + \left(\frac{\partial f}{\partial y}\right)^2 \sigma_y^2, \tag{6.21}$$

$$= \sigma_x^2 + \sigma_y^2, \tag{6.22}$$

$$\sigma_f = \sqrt{\sigma_x^2 + \sigma_y^2}. \tag{6.23}$$

Here we find the result that the variance on f is the sum of variances on x and y. We usually say that errors combined in this way have been **added in quadrature**. This result is independent of the signs of x and y in f, so it is valid for $f = \pm x \pm y$. By following a very similar derivation we are able to show that the relative error on the products or the ratio of x and y is also given by the sum in quadrature of the errors on x and y; however, this is a task left for the reader.

6.2.3 Functions involving powers of x

The final example considered here is that of the function $f = x^n$. If we have determined the value and uncertainty on x, we can use the general formula of

Eq. (6.11) once again to determine the corresponding error on f as

$$\sigma_f^2 = \left(\frac{\partial f}{\partial x}\right)^2 \sigma_x^2, \tag{6.24}$$

$$\sigma_f = nx^{n-1}\sigma_x. \tag{6.25}$$

Here we find a relationship between the uncertainty on f and both the value and uncertainty on x.

6.2.4 Correlations revisited

The concept of correlation between variables was introduced in Section 4.6.2. Just as the determination of one variable may depend on the value of another correlated variable, the error on the measurement of one observable or parameter may depend on the value of another correlated observable or parameter. It is necessary to understand in detail the covariance between correlated variables for such a problem in order to be able to correctly compute errors, or combining repeated measurements of an observable that may be correlated. Returning to Eq. (6.7), we now see that if the covariance between two variables is non-zero that the last term plays an important role in understanding the uncertainty on the combination.

The result in Eq. (6.7) can be generalised for the case where one has an N-dimensional problem. Correlated variables will have terms depending on the covariance of the pairs, and uncorrelated terms will take the form of Eq. (6.11). This generalised result is

$$\sigma_f^2 = \sum_{i,j} \frac{\partial f}{\partial u_i}\frac{\partial f}{\partial u_j}\sigma_{u_iu_j}, \tag{6.26}$$

$$= \sum_i \left(\frac{\partial f}{\partial u_i}\right)^2 \sigma_{u_i}^2 + \sum_{i,j,i\neq j} \frac{\partial f}{\partial u_i}\frac{\partial f}{\partial u_j}\sigma_{u_iu_j}, \tag{6.27}$$

where the first sum corresponds to the uncorrelated error computation, and the second sum corresponds to the correlated error component for some pair of variables u_i and u_j. The indices i and j are over the N dimensions of the problem. The uncorrelated parts form the diagonal of a matrix, and the correlated parts are off diagonal terms. Equation (6.27) can be expressed in matrix form as

$$\sigma_f^2 = \left(\frac{\partial f}{\partial u_1}, \frac{\partial f}{\partial u_2}, \ldots, \frac{\partial f}{\partial u_n}\right) \begin{pmatrix} \sigma_{u_1}^2 & \sigma_{u_1u_2} & & \\ \sigma_{u_1u_2} & \sigma_{u_2}^2 & & \\ & & \ddots & \\ & & & \sigma_{u_n}^2 \end{pmatrix} \begin{pmatrix} \frac{\partial f}{\partial u_1} \\ \frac{\partial f}{\partial u_2} \\ \vdots \\ \frac{\partial f}{\partial u_n} \end{pmatrix}, \tag{6.28}$$

where we see the (symmetric) covariance matrix as the middle term of the right-hand side of Eq. (6.28). This can be re-written in a compact form as

$$\sigma_f^2 = \Delta^T V \Delta, \tag{6.29}$$

where Δ is a column matrix of partial derivatives of f with respect to the u_i and V is the $n \times n$ covariance matrix. Thus if one knows the uncertainties on u_i, covariances between pairs of variables, and is also able to determine the partial derivatives, it is possible to compute the error on f via the appropriate matrix multiplication. The error propagation formalism described here is applied to the problem of tracking a moving object, and determining the error on its trajectory in Section 6.7.2.

6.2.5 Correlated and uncorrelated uncertainties

The preceding discussion raised the issue that some or all of the uncertainties we are interested in may be correlated with each other. If we have some observable x that has a number of sources of uncertainty, all of which are uncorrelated, then we are able to combine the uncertainties in quadrature in order to obtain the total uncertainty on x using a generalisation of the result given in Eq. (6.23).

If, however, the uncertainties on x were correlated, it would be wrong to assume that we can add them in quadrature. Instead correlated uncertainties should be added coherently (or linearly) in order to obtain the total uncertainty. This follows from Eq. (6.27) as can be seen from the following. Given $f = x + y$, where we have determined $\sigma_x = \sigma_y$, and $\rho_{x,y} = 1$, then it follows that the total error on f is given by

$$\sigma_f = \sqrt{\sigma_x^2 + \sigma_y^2 + 2\rho_{x,y}\sigma_x\sigma_y}, \tag{6.30}$$

$$= 2\sigma_x = 2\sigma_y. \tag{6.31}$$

Often a measurement will have sources of uncertainty that are both correlated and un-correlated. In such cases we add the correlated errors linearly, and the un-correlated parts in quadrature. Finally we can consider calculating the total uncertainty by combining the correlated and un-correlated sums in quadrature; however, this implicitly assumes that the uncertainties are all Gaussian in nature. This may not always be the case for systematic uncertainties discussed in Section 6.5.

6.3 Binomial error

It is useful at this point to revisit the binomial probability distribution (Section 5.2) in order to discuss the binomial error and its uses. Let us consider the case where we are trying to measure the efficiency of a detector. Our experiment has two states: the first where we detect what we want to measure, and a second when we don't detect a signal. Knowing in advance that we want to measure the detector efficiency we design our experiment with sufficient redundancy so as to be able to compute this. Then we embark on a detailed study to accumulate data over a period of time in order to estimate the detector efficiency. This efficiency is simply the fraction of detected events normalised by the total number of events measured by a reference system. If our detector is particularly good and fails to detect only one in a thousand events, then the efficiency is $\epsilon = 0.999$. To convince ourselves that we understand the detector efficiency fully we must also determine what the error on ϵ is. We know that the efficiency is

$$\epsilon = \frac{n_d}{n}, \tag{6.32}$$

where n_d is the number of detected events, and n is the total number of trials. The detection of events is a binomial quantity, so the variance is given by Eq. (5.8) and is $n\epsilon(1 - \epsilon)$, where here $p = \epsilon$. From Eq. (6.11) it follows that the uncertainty on ϵ is given by

$$\sigma_\epsilon = \sqrt{\frac{\epsilon(1 - \epsilon)}{n}}. \tag{6.33}$$

This uncertainty is sometimes referred to as the **binomial error** on the efficiency. A practical way to measure the efficiency of a detector for particles that would pass through it is to sandwich that between two other 'trigger' devices, and correlate the detection of some event in both triggers with the presence or absence of an event being detected in the device under study. The underlying reasoning is that if the two triggers are appropriately set up, then one can be certain that, when they both record the passage of a particle, the test device should also have done so. This method can be adapted to the scenario of estimating the efficiency of a set of planar detectors each of unknown efficiency in a stack.

Example. If an efficiency is measured as $\epsilon = 0.999$, from a sample of 1000 events, then the uncertainty on the efficiency is $\sigma_\epsilon = 0.001$, and so the efficiency should be quoted as $\epsilon = 0.999 \pm 0.001$.

If we consider the extreme cases where $\epsilon = 0$ or 100%, then the above formula would suggest that $\sigma_\epsilon = 0$. However, if one thinks about this, then it is unreasonable to expect that a result where an efficiency is trivially good or bad (maximal or minimal) is a perfect measurement. This is a reflection of our ignorance, or lack of information. In order to compute a reasonable estimate of the uncertainty that may be associated with such an efficiency estimate, one must place an upper or lower limit on the efficiency, a subject discussed in Section 7.5.

6.4 Averaging results

6.4.1 Weighted average of a set of measurements for a single observable

If we perform an ensemble of independent measurements to obtain an estimate of the mean and uncertainty of an observable, $x \pm \sigma$, we are free to repeat our experiment in order to accumulate a larger ensemble of data. In such circumstances we can recompute $x \pm \sigma$ for the combined original and new ensembles of data. Usually we don't have the luxury of being able to access all of the data available, for example when previous or competing experiments have made a measurement of which we are performing a more precise cross-check. If this is the case, we need to find a way of combining the two results $x_1 \pm \sigma_1$ and $x_2 \pm \sigma_2$. Naively one might be tempted to average x_1 and x_2 by computing $\bar{x} = (x_1 + x_2)/2$, however, that is not a particularly useful measure of the combination if one of the measurements is more precisely determined than the other. Only in the limit where $\sigma_1 \simeq \sigma_2$ is a good approximation might we consider the arithmetic mean of x_1 and x_2 a good representation of the average value of x.

In order to overcome this problem we can compute a ***weighted average*** derived using a least-squares statistic (see Chapter 9) of two un-correlated measurements, that takes into account the knowledge of uncertainties in each individual measurement using

$$\bar{x} \pm \sigma_x = \frac{x_1/\sigma_1^2 + x_2/\sigma_2^2}{1/\sigma_1^2 + 1/\sigma_2^2} \pm \left(1/\sigma_1^2 + 1/\sigma_2^2\right)^{-1/2}, \tag{6.34}$$

$$= \frac{\sigma_2^2 x_1 + \sigma_1^2 x_2}{\sigma_1^2 + \sigma_2^2} \pm \left(\frac{\sigma_1^2 \sigma_2^2}{\sigma_1^2 + \sigma_2^2}\right)^{1/2}. \tag{6.35}$$

When considering n such measurement this generalises to

$$\bar{x} \pm \sigma_x = \frac{\sum_{i=1}^{n} x_i/\sigma_i^2}{\sum_{i=1}^{n} 1/\sigma_i^2} \pm \left(\sum_{i=1}^{n} 1/\sigma_i^2\right)^{-1/2}. \tag{6.36}$$

Here each measurement is weighted by the reciprocal of its variance in order to compute the average. The corresponding uncertainty is the square root of the inverse of the sum in quadrature of the reciprocals of the variances of the measurements.

6.4.2 Weighted average of a set of measurements for a set of observables

If one wants to compute a weighted average of a set of correlated observables x from a number of measurements M, then the procedure follows a generalisation of the method used for the un-correlated case. For the general case (again obtained using the least-squares method) one obtains a matrix equation relating the covariance matrix V and the set of observables x_j from the jth measurement and the mean value of that measurement set $\bar{x}$. As shown in James (2007), the mean values $\bar{x}$ and covariance matrix obtained for the ensemble of measurements are given by

$$\bar{x} = \left[\sum_{j=1}^{M} V_j^{-1} \right]^{-1} \cdot \left[\sum_{j=1}^{M} V_j^{-1} x_j \right],$$

$$V = \left[\sum_{j=1}^{M} V_j^{-1} \right]^{-1}. \tag{6.37}$$

If we consider the case of M measurements of a set of two observables a and b, then Eq. (6.37) becomes

$$\bar{x} = \left[\sum_{j=1}^{M} \begin{pmatrix} \sigma_a^2 & \sigma_{ab} \\ \sigma_{ab} & \sigma_b^2 \end{pmatrix}_j^{-1} \right]^{-1} \cdot \left[\sum_{j=1}^{M} \begin{pmatrix} \sigma_a^2 & \sigma_{ab} \\ \sigma_{ab} & \sigma_b^2 \end{pmatrix}_j^{-1} \begin{pmatrix} a \\ b \end{pmatrix}_j \right],$$

$$V = \left[\sum_{j=1}^{M} \begin{pmatrix} \sigma_a^2 & \sigma_{ab} \\ \sigma_{ab} & \sigma_b^2 \end{pmatrix}_j^{-1} \right]^{-1}, \tag{6.38}$$

where

$$\begin{pmatrix} \sigma_a^2 & \sigma_{ab} \\ \sigma_{ab} & \sigma_b^2 \end{pmatrix}_j^{-1} = \left[\frac{1}{\sigma_a^2 \sigma_b^2 - \sigma_{ab}^2} \begin{pmatrix} \sigma_b^2 & -\sigma_{ab} \\ -\sigma_{ab} & \sigma_a^2 \end{pmatrix} \right]_j. \tag{6.39}$$

From this one can see that for un-correlated observables in the measurement set we can reduce Eq. (6.37) to a set of calculations for each of the individual observables. On doing this we recover a set of weighted averages of the form given in Eq. (6.36), one for each observable. A worked example of how to compute a weighted average of correlated observables is given in Section 6.7.3.

6.5 Systematic errors and systematic bias

Until now we have considered only errors that are statistical in nature. These are uncertainties on a measurement that result solely from our ability to extract information about an observable assuming that the analysis method is perfect. In reality there are other types of errors that have to be considered when making measurements. These are all called **systematic errors** and are related to uncertainties in our measuring device (i.e. the scale on a meter rule or its equivalent) as well as anything else that may impact upon our result.

Systematic errors should not be confused with a systematic bias on a measurement. What's the difference between bias and error? A **bias** is a systematic offset on a measured observable that results from the procedure used. An error is an uncertainty attributable to an effect. The knowledge of a systematic bias on a measurement also has a systematic uncertainty associated with it. Having determined the value and uncertainty of a bias, one can make a correction to a measured observable to remove the effects of the bias. Such a correction becomes an integral part of the analysis method, and may also provide insight on how to improve the method for future measurements. This can be illustrated through the following example.

Having prepared an experiment to measure the lifetime of cosmic ray muons one makes an assumption that the equipment used by one of your colleagues was sufficiently well calibrated. After taking your data you have a measurement with a relative precision of 1%, but you find that this is 10% smaller than anticipated based on the literature. This difference is significant, so you investigate possible sources of bias in your procedure. Finally the problem is traced back to a poor calibration. In an ideal world at this point you would consider re-running the experiment, discarding your existing data; however, you have a deadline in a few days and only have time to use the data already collected.

In this case, you determine a way that you can quantify the bias from the poor calibration, which corresponds to $-(10 \pm 0.5)\%$ on the measured lifetime. So now you can correct your original value by a shift of $+10\%$, and combine the error on the bias, with the statistical uncertainty for your report. Your final result is compatible with the literature, and the precision of your measurement is 1.5% adding the statistical and systematic errors assuming they may both be correlated, and highlight that one may improve the technique by re-calibrating more often for future measurements.

It is important to note that, in the above illustration, the experimenter finds a problem with the method, and quantifies how this affects the result, prior to correcting the result and including an additional uncertainty from the correction. One thing that must _never_ be done is to 'massage' or fabricate the data! From time to time experiments make more precise measurements of well-known quantities and do find that they are significantly different from those reported in the literature. Such differences are often an indication that the previous method had a systematic effect that was unknown or overlooked and sometimes these will simply be the result of statistical fluctuations. If the people doing those repeat experiments had massaged their data in order to agree with the previous result, they would obtain the wrong results and would have wasted considerable resources in doing so. It is important that you remain an objective scientist when making measurements, and treat surprising results with the same (or more) diligence than you would an expected result.

Sometimes the systematic error associated with the correction will turn out to be small, sometimes it will dominate the total uncertainty reported. Similarly sometimes the bias from an effect will be negligible, and sometimes, as in the above illustration, it will not.

Statistical errors are Gaussian in nature, which is why we add them in quadrature. Systematic errors are by definition almost certainly not. When dealing with systematic uncertainties there is no firm rule on how to combine them. Often one will find that they are combined in quadrature, sometimes they will be added linearly, and the other option is to add some in quadrature and some linearly depending on their nature. Kirkup and Frenkel (2006) discuss systematic uncertainties on measurement for a number of concrete examples. The treatment of systematic errors has to be considered carefully on a case by case basis, and experience will help significantly in determining what to do with them. One definite case is outlined in the following example.

On measuring the probability for a particle to decay into a final state with two neutral pions (π^0), we introduce a systematic uncertainty based on our ability to reconstruct and identify both particles. This uncertainty is typically 3% per π^0 for the BABAR experiment, which was built at the SLAC National Laboratory in California during the late 1990s, and took data between 1999 and 2008. With a $2\pi^0$ final state, we have to add the uncertainty on the reconstruction efficiency for each π^0 linearly because the effect of reconstructing one π^0 is completely correlated with the effect of reconstructing the second. So the systematic uncertainty on measuring such a final state on the BABAR experiment has a total contribution from π^0 reconstruction of 6%. This uncertainty can now be combined in quadrature with any other uncertainties that are considered to be un-correlated with it in order to obtain the total uncertainty on the measurement.

6.6 Blind analysis technique

The concept of blind analysis is not new; for example, Dunnington (1933) measured the value of e/m for the electron this way. This technique has gained popularity in recent years. The previous section discussed the need to remain objective when making a measurement, and to be as diligent as possible so as not to introduce experimenter bias. That is, by favouring one outcome over another the experimenter could unwittingly perform a measurement that enables them to draw the favoured conclusion, irrespective of what the data are actually telling them. The most natural way to unwittingly introduce experimenter bias in a measurement would be to stop investigating possible systematic effects and sources of bias when one has obtained a result in agreement with a previous experiment. This is not a sufficient requirement to ensure that the result you have obtained is correct.

The method behind performing a ***blind analysis*** is to obscure the observable x being measured in an unknown, but repeatable, way such that a measurement can be performed giving the result x_{blind}. In many circumstances, in addition to extracting x_{blind}, it is possible to extract the systematic errors on x_{blind} (and subsequently x) and to perform all necessary cross-checks and validation studies without the need to determine the actual value of the observable being measured. Only once the analysis of data has been completed is the observable 'un-blinded' to give the result $x \pm \sigma_x$, where σ_x is the combined statistical and systematic uncertainty on both x and x_{blind}.

This may seem a rather strange way to perform a measurement; however, it does remove any potential for experimenter bias to influence a result. Of course if a mistake is found in the procedure once the observable has been un-blinded, that mistake has to be rectified, otherwise an incorrect result would be reported. The benefit of performing a blind analysis is that experimenter bias will be absent. The downside of performing such an analysis is that sometimes it is can be impossible or extremely impractical to perform all necessary checks on a particular analysis in order to extract the observable. Where it is possible to perform a blind analysis, the experimenter needs to realistically weigh the advantages against the disadvantages of doing so, including understanding how one can perform any and all necessary cross-checks on the analysis. An example of such a case is one where you search for a particular signal in a sample of limited statistics, and the signal itself depends on a number of parameters, all of which are unknown. How can you perform a blind analysis of something you don't even know exists? If there are a limited number of unknowns it may be feasible to perform a blind analysis when searching for a new effect. When it is not possible to adopt a blind analysis approach, the experimenter needs to rely on remaining objective and ensure that there are sufficient processes in place in an analysis chain to avoid unwittingly biasing the end result. Ultimately, as long as the experimentalist remains objective then the procedures imposed by a blind analysis technique are unnecessary.

6.7 Case studies

This section discusses examples that can be understood by applying some of the techniques discussed in this chapter.

6.7.1 Total uncertainty on a measurement

As described in this chapter, there are two sources of error (statistical and systematic), and understanding how to combine these is an important part of statistical data analysis. The end result will be a measurement comprising a central value and an uncertainty.

Consider the computation of the volume of a crate, where the dimensions of each side are measured to be 1 m using a ruler graduated to half a centimetre. Each length in the x, y, and z direction is therefore measured to a precision of ± 0.25 cm, which is the statistical uncertainty on a measurement. If the ruler is inexpensive and is certified to be accurate to 1 mm over a 1 m length, there will be a systematic uncertainty of ± 1 mm on each length measurement. The total volume of the crate is given by $V = xyz$, where x, y, and z are all $(1.0000 \pm 0.0025(\text{stat.}) \pm 0.0010(\text{syst.}))$ m. The statistical uncertainties can be added in quadrature (see Section 6.2.2) so the total statistical uncertainty on V is 0.0043 m^3. The systematic uncertainty on the measurement of x has the same source as that on y and as that on z, so the systematic uncertainties are all correlated. As a result of this we must add the systematic uncertainties linearly with each other to obtain ± 0.003. Hence the volume of the crate is $(1.000 \pm 0.004(\text{stat.}) \pm 0.003(\text{syst.}))$ m^3. Given that the systematic and statistical uncertainties are un-correlated, one often combines these numbers in quadrature to quote a single total uncertainty. In this case the volume of the crate would be 1.000 ± 0.005 m^3.

While this example may see a little contrived, there are practical uses to such a calculation. For example, the design team at a company specialising in making rulers may consider this level of performance sufficient for their product line of rulers for general use, but insufficient for a line of engineering rulers. They might decide to investigate the implications of being able to make more precise measurements, for example by increasing the number of graduation marks (hence decreasing the total statistical uncertainty on a measurement) and investing in more accurate manufacturing processes to improve the systematic uncertainty on the device made. It is clear from the result obtained, where the uncertainties on a volume measurement made using the existing ruler are $\pm 0.004(\text{stat.}) \pm 0.003(\text{syst.})$, that if one wanted to halve the total uncertainty on the volume measurement both statistical and systematic uncertainties would need to be reduced. The same logical process can be applied when considering more complicated measurement problems.

6.7.2 Tracking uncertainty

One problem that arises in a number of scenarios, is that of tracking an object through space as a function of time. Applications of this problem include tracking airplanes, spaceships, asteroids, comets, charged particles, and so on. Here we consider the simplified case of linear motion, neglecting any corrections to motion resulting from gravity or, in the case of charged particles, a magnetic field. The procedure illustrated by the linear case can be generalised to other tracking problems encountered.

At some time t_0 one can measure the position of an object $\underline{x}_0 = (x, y, z)$ with a certain accuracy resulting in a covariance matrix V. At a slightly later time t_1, a second measurement of the position can be made corresponding to $\underline{x}_1$, and from these measurements it is possible to estimate the velocity of the object being tracked. As velocity is distance travelled over time, this is given by

$$\underline{v} = \frac{\underline{x}_1 - \underline{x}_0}{t_1 - t_0},$$ (6.40)

$$= \frac{\Delta \underline{x}}{\Delta t}.$$ (6.41)

Following on from Eq. (6.29), the variance on the velocity σ_v^2 is given by

$$\sigma_v^2 = \Delta^T \begin{pmatrix} \sigma_{\Delta x}^2 & \sigma_{\Delta x \Delta y} & \sigma_{\Delta x \Delta z} & \sigma_{\Delta x \Delta t} \\ \sigma_{\Delta y \Delta x} & \sigma_{\Delta y}^2 & \sigma_{\Delta y \Delta z} & \sigma_{\Delta y \Delta t} \\ \sigma_{\Delta z \Delta x} & \sigma_{\Delta z \Delta y} & \sigma_{\Delta z}^2 & \sigma_{\Delta z \Delta t} \\ \sigma_{\Delta t \Delta x} & \sigma_{\Delta t \Delta y} & \sigma_{\Delta t \Delta z} & \sigma_{\Delta t}^2 \end{pmatrix} \Delta,$$ (6.42)

where the matrix Δ is given by

$$\Delta = \begin{pmatrix} \frac{\partial v}{\partial \Delta x} \\ \frac{\partial v}{\partial \Delta y} \\ \frac{\partial v}{\partial \Delta z} \\ \frac{\partial v}{\partial \Delta t} \end{pmatrix}.$$ (6.43)

If the time measurements are not correlated with the space measurements, then Eq. (6.42) reduces to

$$\sigma_v^2 = \Delta^T \begin{pmatrix} \sigma_{\Delta x}^2 & \sigma_{\Delta x \Delta y} & \sigma_{\Delta x \Delta z} & 0 \\ \sigma_{\Delta y \Delta x} & \sigma_{\Delta y}^2 & \sigma_{\Delta y \Delta z} & 0 \\ \sigma_{\Delta z \Delta x} & \sigma_{\Delta z \Delta y} & \sigma_{\Delta z}^2 & 0 \\ 0 & 0 & 0 & \sigma_{\Delta t}^2 \end{pmatrix} \Delta$$ (6.44)

$$= \left(\frac{\partial v}{\partial \Delta x}\right)^2 \sigma_{\Delta x}^2 + \left(\frac{\partial v}{\partial \Delta y}\right)^2 \sigma_{\Delta y}^2 + \left(\frac{\partial v}{\partial \Delta z}\right)^2 \sigma_{\Delta z}^2 + \left(\frac{\partial v}{\partial \Delta t}\right)^2 \sigma_{\Delta t}^2$$

$$+ 2\frac{\partial v}{\partial \Delta x}\frac{\partial v}{\partial \Delta y}\sigma_{\Delta x \Delta y} + 2\frac{\partial v}{\partial \Delta x}\frac{\partial v}{\partial \Delta z}\sigma_{\Delta x \Delta z} + 2\frac{\partial v}{\partial \Delta y}\frac{\partial v}{\partial \Delta z}\sigma_{\Delta y \Delta z}$$ (6.45)

$$= \frac{1}{(\Delta t)^2} \left[\sigma_{\Delta x}^2 + \sigma_{\Delta y}^2 + \sigma_{\Delta z}^2 + 2\sigma_{\Delta x \Delta y} + 2\sigma_{\Delta x \Delta z} + 2\sigma_{\Delta y \Delta z}\right]$$

$$+ \frac{|\Delta \underline{x}|^2}{(\Delta t)^4} \sigma_{\Delta t}^2,$$ (6.46)

as the 4×4 covariance matrix in Eq. (6.42) is symmetric. If we assume that the uncertainties on $\underline{x}_0$ and $\underline{x}_1$ are completely correlated, then the uncertainty on Δx is twice that of the uncertainty on a measurement of $\underline{x}$. Assuming that the same is true for the time measurement, one obtains

$$
\begin{pmatrix}
\sigma^2_{\Delta x} & \sigma_{\Delta x \Delta y} & \sigma_{\Delta x \Delta z} \\
\sigma_{\Delta x \Delta y} & \sigma^2_{\Delta y} & \sigma_{\Delta y \Delta z} \\
\sigma_{\Delta x \Delta z} & \sigma_{\Delta y \Delta z} & \sigma^2_{\Delta z}
\end{pmatrix}
= 4
\begin{pmatrix}
\sigma^2_x & \sigma_{xy} & \sigma_{xz} \\
\sigma_{xy} & \sigma^2_y & \sigma_{yz} \\
\sigma_{xz} & \sigma_{yz} & \sigma^2_z
\end{pmatrix}
\qquad (6.47)
$$

(again noting that the covariance matrix is symmetric) and

$$
\sigma^2_{\Delta t} = 4\sigma^2_t. \qquad (6.48)
$$

Using the above covariances with Eq. (6.46) it is possible to compute the error on the velocity, and hence provide a prediction of the trajectory with uncertainties at a given instant in time. The result obtained is

$$
\sigma^2_v = \frac{4}{(\Delta t)^2} \left[\sigma^2_x + \sigma^2_y + \sigma^2_z + 2(\sigma_{xy} + \sigma_{xz} + \sigma_{yz}) \right] + 4\sigma^2_t \frac{|\Delta \underline{x}|^2}{(\Delta t)^4}. \qquad (6.49)
$$

Having determined the form of σ^2_v one is in a position to compute the uncertainty on the velocity, and hence understand what factors dominate the total uncertainty. Given this information it may be possible to propose improvements to the measurement method that may lead to more precise determination of $\underline{v}$ in the future.

6.7.3 Weighted average of measurements of correlated observables

Two particle physics experiments, BABAR at the SLAC National Laboratory in California and Belle at KEK in Japan, have measured the quantities S and C in a number of different decay modes.[1] One of these is in the decay of a neutral B meson into two charged ρ mesons. A non-zero value of S or C would indicate that matter behaved differently from antimatter in a given decay.[2] The two measurements are independent of each other, however S and C are correlated for a given measurement. The results obtained by BABAR and Belle are summarised in Table 6.2, where the BABAR result is more precise than the Belle one. As a result a naive average of the two measurements would not be a good approximation of the average value and one should perform a weighted average.

---

[1] The results reported by Belle used a different notation where $A = -C$, here these have been translated to C.

[2] As an aside, in general there is a small difference between matter and antimatter observed for some decay processes. This is something that is actually required to explain the evolution of our Universe from the Big Bang into the present day matter dominated state. The observed level of matter–antimatter asymmetry to date is approximately one billion times too small to explain the how the Universe has evolved. An entertaining discussion of this topic can be found in the popular book by Quinn and Nir (2007).

Table 6.2 *Measurements of S and C made by* BABAR *and Belle. These measurements are summarised to three decimal places to avoid introducing significant rounding errors in the average, normally when presenting these results one would quote two decimal places for S and C.*

Experiment	S	C	ρ_{SC}	cov_{SC}
BABAR (Aubert *et al.*, 2007)	-0.170 ± 0.207	$+0.010 \pm 0.162$	0.035	-0.0012
Belle (Somov *et al.*, 2007)	$+0.190 \pm 0.310$	-0.160 ± 0.225	0.100	0.0070

The average values of S and C can be obtained by following the procedure outlined in Section 6.4.2 where the measured observables are

$$x_1 = \begin{pmatrix} -0.170 \\ +0.010 \end{pmatrix}, \qquad V_1 = \begin{pmatrix} 0.0430 & -0.0012 \\ -0.0012 & 0.0261 \end{pmatrix}, \qquad (6.50)$$

$$x_2 = \begin{pmatrix} +0.19 \\ -0.16 \end{pmatrix}, \qquad V_2 = \begin{pmatrix} 0.0964 & 0.0070 \\ 0.0070 & 0.0505 \end{pmatrix}, \qquad (6.51)$$

for indices $j = 1, 2$ corresponding to BABAR and Belle, respectively. The sum of the inverse of the covariance matrices is

$$V_1^{-1} + V_2^{-1} = \begin{pmatrix} 33.749 & -0.402 \\ -0.402 & 58.363 \end{pmatrix}, \qquad (6.52)$$

which is used to determine both the central values and covariance matrix for the average of S and C. The inverse of Eq. (6.52) is the covariance matrix of the average, which is given by

$$V = \begin{pmatrix} 0.0296 & 0.0002 \\ 0.0002 & 0.0171 \end{pmatrix}, \qquad (6.53)$$

which can be used to determine the uncertainty on the average values of S and C as well as the correlation between these two parameters. The result of applying the averaging procedure of Eq. (6.37) is

$$\begin{pmatrix} S \\ C \end{pmatrix} = \begin{pmatrix} -0.05 \pm 0.17 \\ -0.06 \pm 0.13 \end{pmatrix}, \qquad (6.54)$$

where $\mathrm{cov}_{SC} = 0.0002$, and hence $\rho_{SC} = 0.009$. If one ignored the correlations between the individual values of S and C for a given measurement then instead of obtaining the correct result of Eq. (6.54) one would have obtained $S = -0.06 \pm 0.17$ and $C = -0.05 \pm 0.13$. This difference is essentially negligible given the uncertainties on these observables and is a consequence of the small value of the correlation between S and C. The other thing worth noting is that the measurement from BABAR is more precise than that from Belle. Therefore one would expect that the average values obtained would be closer to the BABAR values than those from Belle. This is indeed the case, as one can see by comparing the result in Eq. (6.54) with Table 6.2. Having computed a weighted average it is often worthwhile performing this simple cross-check to make sure that the result of the calculation makes sense and to give reassurance that there has not been a mistake in the calculation.

6.8 Summary

The main points introduced in this chapter are listed below.

(1) Understanding the fundamental behaviour of errors, how they are defined, and how one can combine sources of error from different contributions is at the root of all observational and experimental science. The basic principles adhered to for this endeavour are discussed throughout this chapter. One cornerstone of this understanding is the central limit theorem, which leads to the conclusion that statistical errors are Gaussian.

(2) When planning or considering a new experiment that would collect more data than a previous one, it is often possible to extrapolate results from the old experiment, scaling uncertainties by $1/\sqrt{N}$. This assumes that the new experiment will be equivalent to the old one in all respects other than the number of data collected.

(3) A substantial part of this chapter discussed the possibility of combining the uncertainties on measured observables in order to derive some new quantity. This results in Eq. (6.29) for scenarios where one or more of the observables is correlated, and Eq. (6.11) when there are no correlations.

(4) Once a measurement of some observable has been made, this can be incorporated into the knowledge base of the field. From this one can strive to perform more sophisticated experiments to improve on our knowledge of that observable. In doing so we ultimately arrive at a scenario where there are several measurements of some observable, each with differing precision. Given these data, it is possible to compute a weighted average of un-correlated inputs using Eq. (6.36). If the sets of observables from individual measurements are correlated the averaging procedure is modified as discussed in Section 6.4.2.

(5) The concept of systematic effects was discussed in Section 6.5. The associated uncertainty is related to any bias or lack of knowledge in the measurement process. There are no recipes or strict rules to guide an experimenter as to what constitutes a systematic uncertainty; however, through practice one develops an instinct as to what effects may result in a significant systematic uncertainty on a given type of measurement.

(6) The blind analysis technique was discussed in Section 6.6. This technique can be used to remove the possibility of a measurement bias from the experimenter. While this can be useful, it is not always possible to make measurements using this technique, and there is typically an additional effort required to perform a blind analysis.

The Introduction includes a description of how one can measure the acceleration due to gravity g using a simple pendulum. The techniques required to understand in detail how such an experiment can be performed have now all been discussed in earlier parts of this book. You may wish to revisit Section 1.1 in order to review that example and reflect on how your understanding of techniques from this chapter can play a role in improving the measurement method.

Exercises

6.1 What percentage of results would one expect to obtain within $\pm 3\sigma$ of the mean value of an observable when making a set of measurements?

6.2 Compute the weighted average of the un-correlated results 0.655 ± 0.024, 0.59 ± 0.08, and 0.789 ± 0.071.

6.3 A measurement of a complex quantity $\lambda = re^{i\phi}$ has been made, where $r = 1.0 \pm 0.1$, and $\phi = 0.0 \pm 0.2$ (in radians). What are corresponding uncertainties on x and y, the projections of λ onto the real and imaginary axes, respectively?

6.4 What is the uncertainty σ_A on an asymmetry given by $A = (N_1 - N_2)/(N_1 + N_2)$, where $N_1 + N_2 = N$ is the total number of events obtained in the counting experiment.

6.5 The period of oscillation measured using a simple pendulum is found to be $T = (1.62 \pm 0.20)$ s, for a length of (0.646 ± 0.005) m. What is the corresponding value of g obtained from this measurement?

6.6 What is the expected improvement in the precision of g determined if, instead of measuring a single oscillation, one measures 10 oscillations?

6.7 Several measurements of the period of oscillation made using a simple pendulum are $T = 1.62, 1.56, 1.59, 1.50, 1.56, 1.62, 1.59, 1.65, 1.60, 1.56$ s. Given that the length of the pendulum is (0.646 ± 0.005) m, what is the corresponding value of g obtained?

6.8 Compute the weighted average of the two measurements $x_1 = 1.2 \pm 0.3$ and $x_2 = 1.8 \pm 0.3$.

6.9   Compute the weighted average of the two measurements $x_1 = 1.0 \pm 0.5$ and $x_2 = 2.0 \pm 0.5$.

6.10 The ability to *tag* the value of an intrinsic quantity can be useful, however, this process is often associated with a mis-assignment probability given by some parameter ω. The effective tagging efficiency is given by

$$Q = \epsilon(1 - 2\omega), \qquad (6.55)$$

where ϵ is efficiency. Determine the general equation for the uncertainty on Q given values and uncertainties for ϵ, and ω.

6.11 Consider the measurement problem associated with determining the height of a student in a classroom. What possible sources of systematic uncertainty might be associated with that measurement?

6.12 Two experiments have measured the quantities A and B with the following precisions: $A_1 = 2.0 \pm 0.2$, $A_2 = 1.5 \pm 0.5$, $B_1 = 0.0 \pm 0.5$, and $B_2 = -1.0 \pm 0.3$. Given that the first (second) experiment reports a correlation between A and B of $\rho_{AB} = 0.5$ (0.1) compute the weighted average of these results and the covariance between the average values of A and B.

6.13 Two experiments have measured the quantities A and B with the following precisions: $A_1 = 2.0 \pm 0.5$, $A_2 = 1.5 \pm 0.5$, $B_1 = 0.0 \pm 0.1$, and $B_2 = 0.5 \pm 0.2$. Given that the first (second) experiment reports a correlation between A and B of $\rho_{AB} = 0.0$ (0.3) compute the weighted average of these results and the covariance between the average values of A and B.

6.14 Two experiments have measured the quantities A, B, and C with the following precisions: $A_1 = 3 \pm 1$, $A_2 = 2 \pm 1$, $B_1 = 0 \pm 1$, $B_2 = 1 \pm 2$, $C_1 = 0 \pm 1$, and $C_2 = 1 \pm 1$. Both experiments report that the results A, B, and C are uncorrelated with each other. Compute the weighted average for these observables.

6.15 Given the energy–mass relation $E^2 = m_0^2 c^4 + p^2 c^2$, determine the general formula for the variance on the invariant mass m_0 given measurements of energy E and momentum p with precision of 1% each.

7

Confidence intervals

This chapter develops the notion introduced in Chapter 6 on how one defines a statistical error and extends this to look at one- and two-sided intervals (see Sections 7.1 and 7.2). One-sided intervals fall into the categories of upper or lower limits, which we can place on a hypothesised effect or process that has not been observed (see Section 7.2). Each of these concepts can be interpreted in terms of a frequentist or Bayesian approach. Section 9.7.3 discusses the concept of Bayesian upper limits in the context of a fit to data, and Appendix E contains tables of integrals of several common PDFs that can be used to determine confidence intervals and limits.

7.1 Two-sided intervals

The relevance of two-sided confidence intervals is discussed in the context of useful distributions used to represent PDFs. In particular the following sections highlight the use of Gaussian, Poisson, and binomial distributions as particular use cases of such intervals.

In general, for a distribution $f(x)$ given by some variable x, we can define some region of interest in x called a **two-sided interval** such that $x_1 < x < x_2$. If $f(x)$ is a PDF associated with the measurement of an observable x, then the normalised area contained within that interval will correspond to the probability of obtaining a measurement of the observable x that falls within that interval, namely

$$P(x_1 < x < x_2) = \frac{\int\limits_{x_1}^{x_2} f(x)dx}{\int\limits_{-\infty}^{+\infty} f(x)dx}.$$ (7.1)

The denominator is one if $f(x)$ is a properly normalised PDF. The complement $\overline{P(x_1 < x < x_2)} = 1 - P(x_1 < x < x_2)$ is the probability of obtaining a measurement that falls outside this interval. If the PDF is symmetric about the midpoint in the interval, then the probability of obtaining a measurement of x above or below the interval is equal, and corresponds to exactly half of $\overline{P}$. The probability P is referred to as the **confidence level** (CL), or coverage of the interval.

7.2 Upper and lower limit calculations

Upper limits are **one-sided confidence intervals**, with a given CL, where the lower integration limit has been set to $-\infty$ (or a suitable physical bound). We normally quote that some observable is constrained to be less than x_{UL} at a CL of $Y\%$. This means that based on our observations, we are able to deduce a one-sided confidence interval between some lower physical limit and x_{UL} that contains $Y\%$ of the area of the PDF $f(x)$ representing the measurement. The corollary of this is that if the mean value and variance of our measurement are the true mean and variance of the observable under study, then $Y\%$ of the time a repeat measurement will yield a result $x < x_{UL}$. The complement of this probability is the chance that on repeating the measurement we will find a result that satisfies $x \geq x_{UL}$.

The above discussion can be expressed as

$$CL = \frac{\int\limits_{-\infty}^{x_{UL}} f(x)dx}{\int\limits_{-\infty}^{+\infty} f(x)dx}, \tag{7.2}$$

where it is assumed that it is sensible to integrate over all x; this is Eq. (7.1), but with x_{UL} in place of x_2, and the lower limit x_1 set to $-\infty$.

If it is physically uninteresting or inappropriate to integrate over negative values of x, we can consider modifying the limits of integration to obtain

$$CL = \frac{\int\limits_{0}^{x_{UL}} f(x)dx}{\int\limits_{0}^{+\infty} f(x)dx}, \tag{7.3}$$

where we have implicitly multiplied $f(x)$ by some prior $g(x) = $ constant if $x \geq 0$, otherwise $g(x) = 0$. In this approach the PDF $f(x)$ may be a simple distribution or a more sophisticated model such as a χ^2 or likelihood function (see Chapter 9), which is assumed to tend to to zero in the limit $x \to \infty$. Often the implicit step of multiplying $f(x)$ by some prior is glossed over when computing such a limit.

The purpose of placing an upper limit on an observable is to be able to state with some confidence at what level we have ascertained that the effect associated with that observable does not exist. In order for such an approach to be meaningful, we would like the chosen CL to be reliable, while at the same time to not be overly conservative. In practice we often quote upper limits with a 90% or 95% CL. For example, we perform a search for some effect X, and place an upper limit of $x_{UL} < 3 \times 10^{-6}$ at 90% CL on the probability for this effect. Having deduced this we are confident that in 90% of repeated measurements the central value obtained from a subsequent search should be below the obtained upper limit. It follows that in 10% of cases we may find a value that is larger than the quoted upper limit. Here one implicitly assumes that the measurement corresponds to our best estimate of the true value of X. A consequence of this is that often, when searching for an effect, we can find ourselves in the situation where we observe the effect occurring with a probability greater than or equal to a previous upper limit. That previous result is not wrong − it was just statistically unlucky in comparison with the subsequent discovery.

Being able to place such constraints on physical processes is a useful way to express the fact that a sought after effect has not been found, and at the same time to incorporate the sensitivity of the measurement, taking into account measurement uncertainties. The sensitivity of a number counting experiment is the proportion of events that are actually produced and subsequently detected by the experiment. For example, the detector and event reconstruction efficiencies are factors that affect the sensitivity of an experiment. If there are significant systematic uncertainties associated with a measurement, then these should also be incorporated into an upper limit. There are more sophisticated algorithms that can be used in such cases, for example the unified approach discussed in Section 7.6. A brief discussion of a more general Monte Carlo simulated data-based method for computing limits is also discussed (Section 7.7). Sometimes the single event sensitivity (often referred to as SES) is used in order to compare the ability of one experiment to make a measurement relative to another. The SES for a given experiment is simply the total number of events required in order for the experiment to be able to record a single event of interest. For example, the CERN experiment NA62 is designed to accumulate 100 $K^+ \to \pi^+ \nu \bar{\nu}$ events in two years of data taking (NA62 Collaboration, 2010). In order to achieve this, an estimated total of 10^{13} K^+ decays are required. The corresponding SES for this particular measurement at NA62 is therefore one $K^+ \to \pi^+ \nu \bar{\nu}$ event in 10^{11} K^+ decays, i.e. the SES is 10^{-11}.

Lower limits are the lower bound on the extent of an effect, as determined from data. An example of a situation where one might want to place a lower bound on a number is in the computation of an efficiency. If one has a limited number of events to test the efficiency of a detector, and finds that for each of those events the detector is working, then the computed efficiency would be 100%. However, as no detector is perfect, it is desirable to be able to place some lower limit on the computed efficiency of the detector. This is an example of when one would like to compute a limit for a binomial distribution (see Section 7.5). Mathematically a lower limit can be expressed in a similar way to Eq. (7.2) where

$$CL = \frac{\int\limits_{x_{\mathrm{LL}}}^{+\infty} f(x)dx}{\int\limits_{-\infty}^{+\infty} f(x)dx}, \tag{7.4}$$

and x_{LL} is the lower bound on x for the desired confidence level. Where necessary one can restrict the upper limits of the above integral at a physical maximum.

7.3 Limits for a Gaussian distribution

As introduced in Chapter 6, statistical errors are Gaussian in nature, where the Gaussian distribution is

$$G(x, \mu, \sigma) = \frac{1}{\sigma\sqrt{2\pi}} e^{-(x-\mu)^2/2\sigma^2}. \tag{7.5}$$

The mean value of this distribution is given by $x = \mu$, and the variance is σ^2. The fraction of the area contained between $x = \mu - \sigma$ and $x = \mu + \sigma$ (or $|z| < 1$) is 68.3% to three significant figures. This is a two-sided interval, as shown in Figure 7.1 in terms of z.

The Gaussian distribution is symmetric about its mean value, so it follows that given the fraction of area within $\pm 1\sigma$ is 68.3%, then the fraction of area outside of this interval is 31.7%. This is equally distributed above and below the 1σ interval. Hence there is 15.85% of the area to each side of the 1σ interval. By constructing the two-sided interval $\mu \pm \sigma$ we have divided the PDF into three distinct regions. The interpretation of the integrals of these regions falls into the following categories: (i) the probability of events occurring within the interval, (ii) the probability of events occurring below the interval, and (iii) the probability of events occurring above the interval. Naturally there will be many circumstances where we only care if an event lies within or outside of an interval; however, sometimes it is useful to make the distinction if an event occurs above or below an interval. If we consider an experiment where the PDF describing some data is a Gaussian distribution that has been derived from a measurement m, then when we repeat that measurement there would be the expectation that the probability of measuring a value m' within 1σ of the original measurement is 68.3%. Similarly the probability that m' differs from m by more than 1σ is 31.7%.

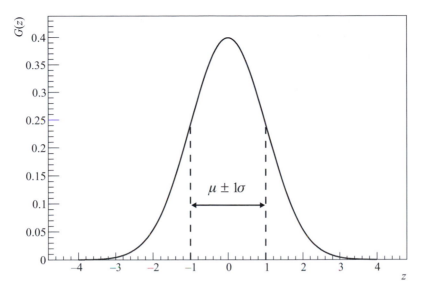

Figure 7.1 A Gaussian distribution indicating the two-sided interval corresponding to $-1 < z < +1$.

We can extend this concept to include asymmetric intervals, where we let one
of the interval limits tend to $\pm\infty$, while the other interval remains finite. In this
case, the interval will split the PDF into two distinct parts where the integrals
represent (i) the probability of events occurring within the interval, and (ii) the
probability of events occurring outside the interval. Figure 7.2 shows a Gaussian
PDF with a one-sided interval where $-\infty < z < +1.64$ ($-\infty < x < +1.64\sigma$).
The probability corresponding to this interval is

$$P = \int_{-\infty}^{+1.64} G(z)dz = \int_{-\infty}^{\mu+1.64\sigma} G(x;\mu,\sigma)dx, \tag{7.6}$$

$$= 0.95, \tag{7.7}$$

given that a Gaussian PDF is normalised to a total integral of unity. So if we make
a measurement of an observable with a mean value of μ and variance σ^2, then 95%
of the time we would expect that a repeat measurement of that observable would
yield a result less than $\mu + 1.64\sigma$. Alternatively we may say that $x \leq \mu + 1.64\sigma$
with 95% CL. The corollary of this is that 5% of the time, the repeat measurement
would yield a result of $x > \mu + 1.64\sigma$.

The importance of one- and two-sided intervals derived from Gaussian PDFs is
the result of the observation, in the large statistics limit, that statistical errors are
Gaussian in nature. Hence an understanding of Gaussian intervals provides a way of

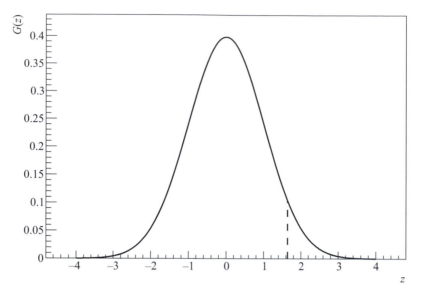

Figure 7.2 A Gaussian distribution indicating the one-sided interval corresponding to $z < 1.64$.

predicting what to expect when an observable is measured, as long as there is some prior indication of what the mean and variance of the distribution should be. Usually this prior indication will be in the form of a previous measurement, or an average of previous measurements (see Chapter 6). One- and two-sided confidence intervals for the Gaussian probability distribution can be found in Tables E.8 and E.9.

7.4 Limits for a Poisson distribution

Rare processes are described by Poisson statistics. In the limit of small numbers, where the Gaussian approximation is invalid, it is useful to consider one- and two-sided intervals of a Poisson distribution to obtain intervals and upper limits. The following discussion mirrors the previous section. Two significant differences between the Poisson and Gaussian distributions are that r, unlike x, is a discrete parameter and the Poisson distribution

$$f(r, \lambda) = \frac{\lambda^r e^{-\lambda}}{r!}, \tag{7.8}$$

is not symmetric about its mean value λ (see Figure 5.4), whereas a Gaussian distribution is. This fact is relevant when constructing a two-sided interval and in particular when determining the $\pm 1\sigma$ uncertainty interval on a measured observable in the limit of small statistics. Such a two-sided interval can be constructed by integrating the Poisson distribution for a given r such that the limits λ_1 and λ_2 are equally probable in order to obtain the desired CL. In doing so we naturally determine an

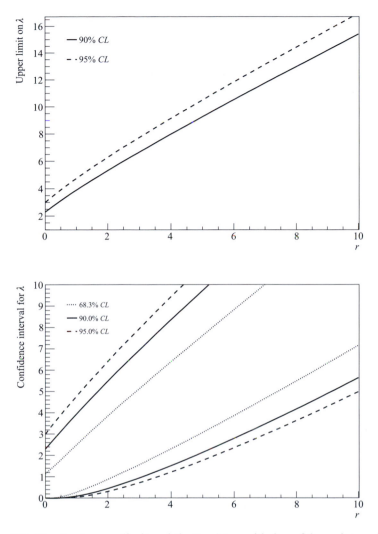

Figure 7.3 The (top) upper limit and (bottom) two-sided confidence intervals for a Poisson distribution of r observed events for (dotted) 68.3%, (solid) 90%, and (dashed) 95% confidence levels. The allowed region for two-sided confidence intervals is between the curves drawn for a given CL.

asymmetric interval about the mean value λ. If we are measuring some quantity where we wish to express 68.3% coverage about a mean, then an asymmetric PDF such as the Poisson distribution naturally leads to an asymmetric uncertainty. As with the Gaussian case, a one-sided interval is a matter of integrating $f(r, \lambda)$ for a given observed number of events r, to obtain a limit with the desired coverage.

Figure 7.3 shows the one- and two-sided confidence intervals obtained for λ in a counting experiment as a function of the number of observed signal events r. The upper limit is quoted in terms of both 90% and 95% CL, as these are commonly found as the levels of coverage used in many scientific publications. The corresponding two-sided interval plot also includes the 68.3% CL contours in order to be able to enable a comparison with the Gaussian σ. One-sided integral tables of the Poisson PDF can be found in Appendix E. The case study described in Section 7.8.2 gives an example of using a Poisson distribution to set a confidence interval. While these intervals are represented by a smooth distribution, one should note that the possible outcomes of an experiment are in terms of discrete numbers of events.

The situation encountered where one has a non-zero background component modifies the previous discussion on computing limits. For such a scenario, where one observes N_{sig} signal events and N_{bg} background events, both of which are distributed according to a Poisson distribution with means λ_{sig} and λ_{bg}, respectively. One can show that the sum of these two components is also a Poisson distribution with a mean of $\lambda_{sig} + \lambda_{bg}$. Given sufficient knowledge of λ_{bg}, one can proceed to compute limits on λ_{sig}. This situation is discussed in Cowan (1998), which also highlights issues surrounding measurements involving large backgrounds with small numbers of observed signal events. This particular problem is also discussed in the context of the unified approach in Section 7.6.

7.5 Limits for a binomial distribution

Setting a confidence interval on some binomial quantity can be done in an analogous way as described above for the Poisson distribution. The likelihood for a binomial distribution is shown in Figure 5.2, with a number of trials N, of which r are successful, and the probability of success or failure is given by p or $1 - p$, respectively. Given an experiment consisting of N trials with r successes one can obtain limits on the allowed values for p by integrating the likelihood distribution in order to obtain the desired level of coverage. Figure 7.4 illustrates this for a binomial experiment with 10 trials, of which two were successful. The upper limit obtained on p at 95% CL is 0.47. Unlike the Poisson distribution, the binomial distribution has a discrete number of trials N and a discrete number of successes r in the outcome, so it is not possible to produce two-dimensional contours of one- and two-sided limits on p scanning through r for a given N. However, for a particular scenario one can perform the appropriate one- and two-sided integrals of distributions, such as that shown, in order to obtain the corresponding confidence intervals.

The concept of a binomial error was introduced in Chapter 6 via measurement of detection efficiency. A pathology was highlighted when the efficiency was either 0% or 100%. For these scenarios one is either minimally or maximally efficient and the binomial error given by Eq. (6.33) is determined to be zero. It is now possible to revisit this situation to consider the appropriate outcome of an extreme scenario.

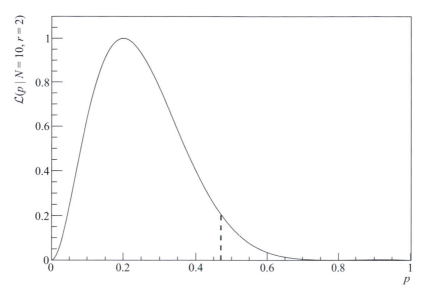

Figure 7.4 The likelihood distribution as a function of p for a binomial experiment with 10 trials, of which two were successful. The dashed vertical line corresponds to the 95% CL upper limit.

As discussed in Chapter 5, one can compute a likelihood distribution for a given outcome of a binomial experiment. This distribution (for example see Figures 5.2 and 7.4) is computed for a given number of trials with a given number of successes, and from this one can determine the relative likelihood for p, the probability of success. In terms of the binomial error p is replaced by the detection efficiency ϵ and the resulting likelihood distribution $\mathcal{L}(\epsilon|N, r)$ can be used to determine a lower (or upper) limit on detection efficiency.

Example. One thousand events are recorded in order to determine the detection efficiency of a silicon detector. In each event the silicon detector produced a clear signal, indicating that there was no recorded inefficiency. Determine a lower limit on the efficiency of this device.

In order to compute the lower limit one must first determine the likelihood distribution $\mathcal{L}(\epsilon|1000, 1000) = \epsilon^{1000}$. This distribution can be integrated from some lower limit ϵ_{LL} to one in order to obtain the desired level of coverage. As a result of this integration the detector is found to be $> 0.9977\%$ efficient at 90% CL (see Figure 7.5).

7.6 Unified approach to analysis of small signals

A problem that can be encountered when dealing with small signals is that the constructed intervals can sometimes be an empty set. To avoid such problems

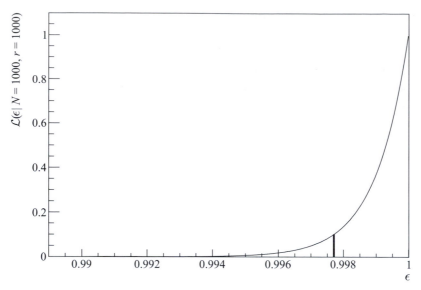

Figure 7.5 The likelihood distribution as a function of ϵ (detection efficiency) for a sample of 1000 events, all of which were detected. The vertical line corresponds to the 90% CL lower limit.

Feldman and Cousins (1998) proposed a so-called unified approach to obtain a frequentist confidence interval. In some fields this is referred to as the *Feldman–Cousins method*. In this method one constructs an interval $[\mu_1, \mu_2]$ allowed for some parameter μ, given a measurement of the observable x, using a likelihood ratio based ordering principle. The probability that the true value of the parameter given by μ_t is contained within the interval is denoted by the confidence level α, i.e. for some arbitrary PDF

$$P(\mu \in [\mu_1, \mu_2]) = \alpha. \tag{7.9}$$

The unified approach to constructing the confidence interval introduces an ordering principle, where values of μ are sequentially added to the interval in order of a likelihood ranking, stopping when the desired coverage is obtained. The likelihood ranking is obtained by normalising the probability $P(x|\mu)$ for a given value of μ by the probability $P(x|\mu_{\text{best}})$ for the largest physically allowed estimate of the the parameter μ, called μ_{best}. For each value of μ one can compute the likelihood ratio

$$R = \frac{P(x|\mu)}{P(x|\mu_{\text{best}})}. \tag{7.10}$$

As with other frequentist interval estimations, discrete problems will tend to have intervals that suffer from incorrect coverage; however, by construction this method either has correct coverage or suffers from over coverage. Other pathologies exist for implementations of this method and are discussed in Feldman and Cousins (1998).

To illustrate how the likelihood ratio ordering algorithm works in practice, the following considers the scenario encountered when attempting to compute a confidence interval for a Poisson observable, resulting from some measurement where one has observed N events with an expectation of N_b background events. Here the parameter μ corresponds to the mean number of signal events that can be inferred from the experiment. The Poisson probability for this scenario is given by

$$P(N|\mu) = \frac{(\mu + N_b)^N e^{-(\mu + N_b)}}{N!}, \tag{7.11}$$

as can be seen from Eq. (5.23), where here $\lambda = \mu + N_b$, and $r = N$. The value of μ_{best} is given by

$$\mu_{\text{best}} = \max(0, N - N_b), \tag{7.12}$$

which is always physical, and represents the most probable value of the signal yield. For a number of observed events N less than the expected number of background events, one takes the most probable value of signal μ_{best} to be zero. If $N > N_b$, then $\mu_{\text{best}} = N - N_b$. It follows from Eqs. (7.10) and (7.11) that

$$R = \frac{(\mu + N_b)^N e^{-(\mu + N_b)}}{(\mu_{\text{best}} + N_b)^N e^{-(\mu_{\text{best}} + N_b)}}. \tag{7.13}$$

Given this information it is possible to construct a two dimensional confidence interval as a function of μ and N for a given value of N_b via a two step process. The first step is to construct one-dimensional intervals for a given μ and fixed N_b over a suitably large range of N. Having constructed an interval in N, the second step is to repeat the process for different values of μ. The set of one-dimensional intervals can be used to create a two-dimensional interval. Equation (7.12) sets the value of μ_{best}, which will change for each assumed value of N, and for a given combination of μ, N, μ_{best} and N_b one can compute $P(N|\mu)$, $P(N|\mu_{\text{best}})$ and the ratio R. By ranking from most to least probable outcome and then summing up the values of N falling within the desired coverage one can obtain a confidence band in the $N - \mu$ plane. This band corresponds to an interval in N with the desired coverage for the number of observed events for some true value of μ and N_b that one would expect to find on making a measurement. While this information is useful when planning an experimental search (if one builds an experiment to look for some effect, then it is useful to know how many events might be obtained), it is of little use once you have performed a measurement and obtained some number of events $N = N_{obs}$. In this second scenario the process of building the two-dimensional interval continues as one can iterate through different possible values of μ in order to determine a

Table 7.1 *Computations used to construct the 90% C L interval in N given an expected background of three ($N_b = 3$) for an expected number of signal events $\mu_{(best)} = 2$. The final column indicates the cumulative coverage of the confidence level constructed by including all values of N (in order of decreasing R) from 1 down to the value of that particular entry.*

N	μ_{best}	$P(N|\mu)$	$P(N|\mu_{best})$	R	Rank	In interval	CL
0	0	0.0067	0.0498	0.1353	11	no	0.9863
1	0	0.0337	0.1494	0.2256	9	no	0.9615
2	0	0.0842	0.2240	0.3759	7	yes	0.8915
3	0	0.1404	0.2240	0.6266	5	yes	0.7420
4	1	0.1755	0.1954	0.8981	3	yes	0.4972
5	2	0.1755	0.1755	1.0000	1	yes	0.1755
6	3	0.1462	0.1606	0.9103	2	yes	0.3217
7	4	0.1044	0.1490	0.7010	4	yes	0.6016
8	5	0.0653	0.1396	0.4677	6	yes	0.8073
9	6	0.0363	0.1318	0.2752	8	yes	0.9278
10	7	0.0181	0.1251	0.1449	10	no	0.9796

set of confidence intervals. Together this set provides us with the two-dimensional confidence interval in the $N - \mu$ plane. There are two pathologies manifest as a result of this process. The first is that the desired coverage is almost never obtained – for most assumed values of μ this method will over-cover. While not ideal, this situation is an improvement over traditional approaches that can lead to both under and over coverage. Secondly the confidence interval has non-singular values due to the discrete nature of the Poisson distribution. The following example illustrates the process of using the ordering principle to construct a 90% CL interval in N for given values of μ and N_b.

Example. Consider the scenario where one expects to observe $\mu = 2$ signal events with a background $N_b = 3$ in a number counting experiment. What is the 90% CL on N for this? In other words, with 90% confidence how many events will this experiment actually observe? In order to answer this question one needs to construct a confidence band for $\mu = 2$ and $N_b = 3$, which could be used as a starting point to calculate a two-dimensional confidence region, if one so desired.

For a given value of N the value of μ_{best} is given by Eq. (7.12), and both $P(\mu, N)$ and $P(\mu_{best}, N)$ are determined using Eq. (7.11). Using these one can compute R as shown in Table 7.1. The most likely outcome corresponds to the case where one obtains five events based on the expectation of $\mu = 2$ and $N_b = 3$, and this forms the seed about which one constructs an interval. One can see from the table that the probability for this outcome to occur is 17.5%, which is quite large, and $R = 1$ by construction.

It is interesting to note that on applying the ordering principle for this method one can see that the $N = 6$ outcome is preferred over the $N = 4$ outcome even though $P(6, 2) < P(4, 2)$. The underlying principle of this method is not to consider the probability of a given outcome in isolation, or compared against some other outcome, but to interpret a given outcome in terms of the most probable physical outcome given those same circumstances (i.e. with respect to $P(N, \mu_{\text{best}})$). The 90% CL interval constructed in this example is the interval $N \in [2 - 9]$ events. The coverage corresponding to this interval is actually 92.78%, larger than the required value of 90%.

7.7 Monte Carlo method

For complicated PDFs it is often convenient to simulate the measurement process using a Monte Carlo (MC) based event simulation. The MC simulation will use a sequence of pseudo-random numbers to generate simulated data based on a model, and that simulated data can be used to place limits on an observable based on an assumed input (e.g. the result of a measurement). Using such a technique one can often generate a large number of simulated data in order to test the measurement procedure. The estimate of an upper limit at some confidence level CL can be determined via

$$CL = \frac{\sum_{i=1}^{x=x_{UL}} y_i x_i}{\sum_{i=1}^{N} y_i x_i}, \tag{7.14}$$

where the sum is over the binned data. The x value of the ith bin is x_i, and there are y_i entries in that bin, where the value of the upper limit is given by $x = x_{UL}$. Often one finds that it is necessary to interpolate between two adjacent bins of data in order to obtain an estimate that corresponds to the desired confidence level instead of reporting a limit that results in under or over coverage.

A confidence level that lies between bins i and $i + 1$ can be estimated assuming that the confidence level varies linearly in x across the two bins. The sum over the data up to and including the ith bin is given by S_i, thus the probability contained within the interval i to $i + 1$ is $S_{i+1} - S_i$. Assuming a linear approximation to the change in probability as a function of x, the fraction

$$f = \frac{CL - S_i}{S_{i+1} - S_i}, \tag{7.15}$$

can be used to estimate the x value of the desired confidence level, i.e.

$$x_{UL} = x_i + f(x_{i+1} - x_i). \tag{7.16}$$

In general this is an issue for frequentist limit calculations. One should check the coverage obtained for a limit to ensure that the difference between the coverage quoted and the target coverage is understood, and where appropriate one should adjust the derived limit to regain the correct level of coverage (see Section 7.8.3), or report the actual CL obtained.

7.8 Case studies

This section discusses a few case studies that can be understood by developing some of the concepts and applying some of the techniques discussed in this chapter and earlier parts of the book.

7.8.1 Multivariate normal distribution

The multivariate normal distribution is an extension of a one-dimensional Gaussian to an n-dimensional space, where in general one allows for a non-trivial covariance between any combination of pairs of dimensions. This distribution can be used to describe uncertainties (or more precisely confidence regions) obtained when measuring a set of quantities that are correlated with each other, and is often used in the bivariate case ($n = 2$) to illustrate so-called error ellipses in a two-dimensional plane. These ellipses are contours of equal confidence level values (usually 68.3% corresponding to a 1σ interval, or some multiple $N\sigma$). The bivariate normal distribution is obtained when allowing for x and y to be correlated and this is given by

$$P(\underline{x}) = \frac{1}{2\pi |V|^{1/2}} e^{-(\underline{x}-\underline{\mu})^T V^{-1}(\underline{x}-\underline{\mu})/2}, \tag{7.17}$$

where here V is a 2×2 covariance matrix, $\underline{x}$ is the column matrix with elements x and y, and $\underline{\mu}$ is the corresponding column matrix of the means of x and y. The matrices V and V^{-1} can be written in full as

$$V = \begin{pmatrix} \sigma_x^2 & \sigma_{xy} \\ \sigma_{xy} & \sigma_y^2 \end{pmatrix}, \tag{7.18}$$

and

$$V^{-1} = \frac{1}{1-\rho^2} \begin{pmatrix} 1/\sigma_x^2 & -\rho/\sigma_x\sigma_y \\ -\rho/\sigma_x\sigma_y & 1/\sigma_y^2 \end{pmatrix}, \tag{7.19}$$

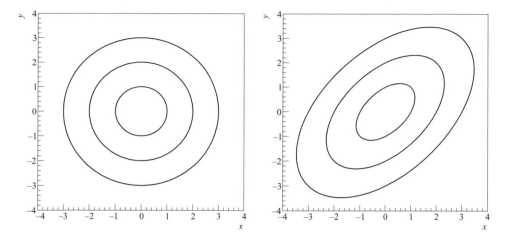

Figure 7.6 The 1σ, 2σ, and 3σ contours for bivariate normal distributions for (left) uncorrelated and (right) correlated parameters x and y. The covariance assumed for the plot on the right is $\sigma_{xy} = 0.5$ ($\rho_{xy} = 0.5$).

where ρ is the Pearson correlation given by Eq. (4.23). Hence the bivariate normal distribution can be written in full as

$$P(x, y) = \frac{1}{2\pi\sigma_x\sigma_y\sqrt{1 - \rho^2}} \tag{7.20}$$

$$\times \exp\left[\frac{-1}{2(1 - \rho^2)}\left(\frac{(x - \mu_x)^2}{\sigma_x^2} - \frac{2\rho(x - \mu_x)(y - \mu_y)}{\sigma_x\sigma_y} + \frac{(y - \mu_y)^2}{\sigma_y^2}\right)\right].$$

One can see how this relates to the simplified case of two uncorrelated variables x and y, in the limit where $\sigma_{xy} = 0$ ($\rho = 0$), and both V and V^{-1} are diagonal matrices. In this case the distribution in Eq. (7.20) reduces to

$$P(x, y) = \frac{1}{2\pi\sigma_x\sigma_y}e^{-(x-\mu_x)^2/2\sigma_x^2+(y-\mu_y)^2/2\sigma_y^2}, \tag{7.21}$$

which follows directly from the discussion of the Gaussian distribution in Section 5.4. Figure 7.6 shows 1σ, 2σ, and 3σ contours for a bivariate normal distribution where x and y are uncorrelated i.e. Eq. (7.21) as well as a distribution where x and y are correlated i.e. Eq. (7.20). In the special case of $\sigma_{xy} = 0$ and $\sigma_x^2 = \sigma_y^2 = 1$ shown in the figure, the error ellipses are concentric circles. If, however, there is a non-trivial correlation then these ellipses deviate from circles and the major and minor axes of the ellipses are not collinear with the horizontal and vertical axes.

Confidence regions shown in the figure have mean values $\mu_x = \mu_y = 0$. The innermost contour corresponds to a 1σ confidence level, and the successive outer contours correspond to 2σ and 3σ confidence levels, respectively. If these distributions corresponded to the PDFs related to some measurement of the parameters x and y, then a measurement within the innermost contour would be compatible with $(x, y) = (0, 0)$ at a level better than 1σ. A measurement falling within the inner two contours would be compatible with $(x, y) = (0, 0)$ at a level better than 2σ, and so on. If the parameters x and y are correlated, then the confidence level shape changes as can be seen by comparing the left- and right-handed plots in Figure 7.6. If one neglected the correlation between two variables then, as shown in the figure, it is possible that reported confidence regions would be quite different from the actual confidence regions.

Given a two-dimensional probability density function $P(x, y)$ representing a measurement of two observables x and y, one can integrate over one of the observables in order to obtain some marginal probability density function $P(x)$ or $P(y)$. If $P(x, y) = P(x)P(y)$ this is a straightforward procedure. Otherwise one can choose to use numerical or analytic means to perform the integral depending on the functional form of $P(x, y)$. In general for a bivariate normal distribution $\rho \neq 0$ and hence it is not possible to express $P(x, y)$ as the product of two independent PDFs. It is possible to simplify the problem by performing a transformation of the data from the correlated (x, y) basis to an un-correlated (u, v) one as discussed in Section 4.6.4. After transforming the data to the (u, v) basis the bivariate normal distribution would simplify to the form given in Eq. (7.21).

The multivariate normal distribution is a simple extension of Eq. (7.17) where

$$P(\underline{x}) = \frac{1}{(2\pi)^{n/2}|V|^{1/2}} e^{-(\underline{x}-\underline{\mu})^T V^{-1}(\underline{x}-\underline{\mu})/2}, \tag{7.22}$$

and now the column matrices are read as having n elements, and V and V^{-1} are the $n \times n$ covariance matrix and its inverse.

7.8.2 *Upper limit calculation from a search*

The T2K experiment has been constructed in Japan with the goal of searching for one type of almost massless neutral particle, called a neutrino, to change into another type. In all there are three types of neutrino: the electron neutrino ν_e, the muon neutrino ν_μ, and the tau neutrino ν_τ. The result of the first search performed at this experiment for ν_μs to change into ν_es is reported by Abe *et al.* (2011). The T2K Collaboration determined a background expectation of 1.5 ± 0.3 events, and observed six candidate $\nu_\mu \rightarrow \nu_e$ events. The the uncertainty ± 0.3 quoted here is the systematic uncertainty on the background estimation. From this result T2K conclude that the probability to observe six or more events given the expected level of background is 7×10^{-3} (equivalent to 2.5σ significance in terms of a Gaussian

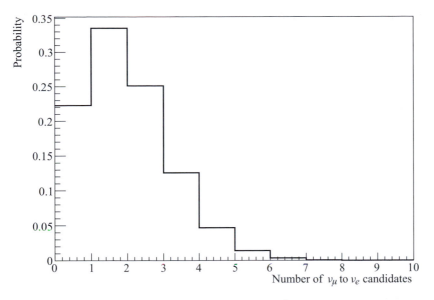

Figure 7.7 A Poisson distribution generated using 10^6 simulated experiment with a mean value of $\lambda = 1.5$.

uncertainty). The details of the calculation performed by this experiment can be found in the reference, and the following provides an illustration of how one might compute this probability using both frequentist and Bayesian methods.

A possible frequentist method

One can consider performing an ensemble of Monte Carlo simulated pseudo-experiments, where one simulates a Poisson distribution with a mean $\lambda = 1.5$ corresponding to the mean expected background. If one does this N times, then the probability can be computed (neglecting the systematic uncertainty of 0.3 events) by determining the fraction of simulated experiments that result in six or more events appearing. As N tends to infinity, the probability obtained tends to the true value of the six events being the result of a background fluctuation. This situation is illustrated in Figure 7.7, where from 10^6 Monte Carlo simulated experiments, one finds a probability of observing six or more events, with an expectation of 1.5 background to be 0.45%. One can take into account the systematic uncertainty on the background in a simple way by shifting the expected background level by ± 0.3 events and repeating the ensemble of Monte Carlo simulated pseudo-experiments, or if one assumes that the quoted systematic uncertainty is Gaussian in nature, then it would be possible to convolute a Poisson distribution with $\lambda = 1.5$ events with a Gaussian distribution of width 0.3 events, and use the resulting distribution to simulate the pseudo-experiments. It should be noted that for this example, one could also numerically integrate the Poisson distribution given in Eq. (5.23) in order to obtain the corresponding probability of observing five or less events, which is discussed in Section 7.4. The complement of this would provide the probability of observing six or more.

A more robust frequentist method would be to adopt the unified approach described in Section 7.6, which discusses this particular problem.

A possible Bayesian method

Given that the background events are Poisson in nature, one can compute a Bayesian probability by defining a prior $P(B)$ for the number of events observed (e.g. a uniform prior for physically allowed values of the observed number of events), and then multiplying a Poisson distribution by the prior to obtain $P(A|B)P(B)$. One can then compute the probability from the following integral ratio

$$1 - P = 1 - \frac{\int_0^6 P(A|B)P(B)dA}{\int_0^\infty P(A)dA} = \frac{\int_6^{+\infty} P(A|B)P(B)dA}{\int_0^\infty P(A)dA}, \quad (7.23)$$

where the lower limit is given by the physical bound of observing no events. This integral ratio depends on the mean value of λ used for the Poisson distribution. Systematic uncertainties can be included in the calculation in the same way as discussed for the frequentist method. Having computed the probability, one should check the dependence of the result obtained on the choice of the prior probability used.

The situation of placing a limit on, or determining the p-value of, a signal yield observed in a sample of data is useful when one wants to address the question of whether something is or is not happening. However, often one is interested in translating an observed yield into a probability that includes detection efficiency and other factors, i.e. taking into account the sensitivity of an experiment (with associated uncertainties). Cousins and Highland (1992) discuss how to approach this particular problem.

7.8.3 Coverage: when is a confidence level not a confidence level?

In order to correctly report a confidence interval one has to understand the coverage associated with that interval. If one considers a one-dimensional problem, there are three pieces of information required in order to quote an interval: (i) the confidence level (or coverage) of the interval, (ii) the lower limit, and (iii) the upper limit. By construction, a Bayesian limit has the correct coverage for example if one considers the Poisson limit problem discussed above then this is implied by Eq. (7.23). A frequentist method, however, does not necessarily report the desired coverage, as can clearly be seen from the discussion of the unified approach (Section 7.6). The discrete nature of the Poisson distribution when applying the unified method leads to a limit that either reports the correct coverage or a more conservative one. This is considered a virtue of the method by the authors. But if one wants to compute a given coverage, for example a 90% CL upper limit, then the discrete nature of that problem generally results in the practitioner choosing which limit to quote: do they quote a 90% CL and report a limit that has a greater than 90% coverage, or do they report the actual coverage obtained and in doing so lose the ease with which their result may be understood in comparison with other experiments? Often the solution taken is the former, with the argument that it is good to be conservative. Formally, however, this could lead to biased interpretation of results as you have mis-represented the precision of your experiment, which is clearly not good practice.

If you are faced with the problem of trying to report a confidence interval where the coverage does not correspond to some 'standard' level, e.g. 68.3%, then it is better to stipulate the coverage obtained as accurately as possible so that someone trying to interpret your result is able to do so without any ambiguity or mistake. If you fall foul of the temptation to quote a limit that under or over covers, then your result may be interpreted in a way that is not correct, and in extreme circumstances could even create some level of confusion with that particular measurement. Just as Bayesian statisticians are concerned about prior independence of the results obtained from data analysis, so frequentists should be concerned about the coverage of their limits or intervals.

7.9 Summary

The main points introduced in this chapter are listed below.

(1) The notions of confidence levels (CL) and confidence intervals as extensions of the definition of an uncertainty are discussed in Section 7.1. The coverage or CL of an interval may take an arbitrary value, and is not restricted by the normal convention for $\pm 1\sigma$ of a Gaussian interval corresponding to a probability of 68.3%.

(2) Both one- and two-sided confidence intervals can be defined, where one-sided intervals can be used to place upper or lower bounds on effects occurring or not. Two-sided bounds can be used to constrain our knowledge of an observable determined from a measurement. In order to fully appreciate how to make the

decision that some effect may or may not exist, one should review the basic concepts of hypothesis testing as discussed in Chapter 8.

(3) The concepts of upper and lower limits as a special cases of a one-sided confidence interval is discussed in Section 7.2.

(4) While analytic integrals of PDFs are often sufficient to determine confidence intervals, for complicated problems it is sometimes more practical to use Monte Carlo techniques. This possibility is discussed in Section 7.7.

(5) It may be necessary to check the coverage of a frequentist upper limit to ensure that the stated confidence level is reproduced accurately. Often one finds that there is over or under coverage for a given limit, and care should be taken to correct for this where appropriate.

Exercises

7.1   Determine the two-sided Gaussian intervals with a coverage of 50%, 75%, 90%, and 99%.

7.2   If an observable is distributed according to a Gaussian PDF with a mean and standard deviation of $\mu = 1.0$, and $\sigma = 1.5$, respectively, what is the 90% confidence level upper limit on the observable?

7.3   Determine the 90% confidence level upper limit on some observable x given the measurement $x = 0.5 \pm 0.5$, where the physically allowed range is given by $x \geq 0$. Assume that the uncertainty on x is distributed according to a Gaussian PDF.

7.4   Consider the process of a particle decaying into an extremely rare final state, one that has yet to be observed by measurement: $A \rightarrow X$. If you perform a search for this decay, and observe one event, estimate the 95% confidence level upper limit on the probability of $A \rightarrow X$ occurring (in number of events).

7.5   Given $x = 1.0 \pm 0.7$, compute the 90% confidence level upper limit on x, assuming that the error is Gaussian. Compute the change in upper limit obtained if you also consider an additional systematic error of 0.4 (also Gaussian in nature).

7.6   Given that three events are observed for a Poisson process, estimate the value and uncertainty on λ.

7.7   Compare the 95% CL upper limit obtained on λ for three observed events with the two-sided interval obtained for the previous question. What conclusion do you make?

7.8 For an unbiased coin tossed ten times, what is the 90% CL upper limit on the number of heads expected?

7.9 Determine the 90% CL upper limit on p for a binomial distribution with ten trials, none of which are successful.

7.10 What is the 90% CL lower limit on the efficiency of a detector that has successfully detected 100 out of 100 events?

7.11 Given a PDF of e^{-x} describing some quantity that is physical for $x \geq 0$, compute the Bayesian 90% CL upper limit on x.

7.12 What is the 95% Poisson upper limit obtained on λ having observed no events in data?

7.13 The parameter S is measured to be -1.2 ± 0.3. What is the Bayesian 90% CL upper limit on S given that S is physically bound within in interval $[-1, 1]$?

7.14 Determine the 90% CL upper limit on the signal yield μ given a background expectation of one event and an observed yield of one event for a rare decay search (note: this problem can be numerically solved once the initial integrals have been performed).

7.15 Use the unified approach to determine the 90% CL interval for N given an expectation of two signal and two background events for a rare process.

8

Hypothesis testing

8.1 Formulating a hypothesis

Up until now we have discussed how to define a measurement in terms of a central value, uncertainties, and units, as well as how to extend these concepts to encompass confidence levels (both one- and two-sided). A related aspect of performing a measurement is to test a theory or, as it is usually phrased, a *hypothesis*. For example, consider the case where a theorist writes a paper proposing the existence of some effect that can be tested via some physical process. It is then down to an experimentalist to develop a method that can be used to test the validity of that theory. In this example the default hypothesis (usually referred to as the *null hypothesis* and often denoted by H_0) would be that the theory is valid, and the experimenter would then embark on a measurement that could be used to test the null hypothesis.

Having defined a null hypothesis, by default the complement of that hypothesis exists as an *alternative hypothesis* often denoted by H_1. Given data we can test the null hypothesis to see if the data are in agreement or in disagreement with it. In order to qualify what is meant by data being in agreement with the null hypothesis, we need to quantify what we consider to be agreement (at some CL). There are four possible outcomes to any given test with data: the null hypothesis can be compatible with or incompatible with the data, and the data can in reality either be based on the null hypothesis or not. If an effect is real, then we can make a measurement, and compare the results of that measurement with the null hypothesis. Some fraction of the time we do this, we will observe that the data agree with the experiment at some pre-defined confidence level. Thus some fraction of experiments where we do everything correctly we will obtain the wrong conclusion. A mis-classification of this kind is called a *type I error* and the probability of obtaining a type I error is often denoted by α. The opposite can also be true – we can make a measurement to test that the null hypothesis is wrong, which is in reality the truth. On performing a correct measurement we can indeed find that the data support that there is no effect. Some fraction of the time, however, we will naturally reach the wrong conclusion, concluding that a non-existent hypothesis is actually true. This type of mistake is called a ***type II error***, which is often denoted by β. The rest of this chapter discusses such tests of the null hypothesis.

8.2 Testing if the hypothesis agrees with data

In order to determine if a hypothesis is compatible with data we must first specify to what degree of belief we aim to make a statement, or alternatively what is our acceptable level of drawing a wrong conclusion. If we want to be certain at the 90% CL that we are able to establish an effect proposed, then by definition 10% of the time we would expect to make a measurement of an effect that was real, and draw the conclusion that the data and theory were incompatible. This corresponds to a type I error rate $\alpha = 10\%$. If we incorrectly reject the hypothesis of a real effect we call this a ***false negative***. As the consequence of drawing such a conclusion can be quite profound, we must be precise and state what criteria are used to draw our conclusions.

Consider the following example: that of an ensemble of Monte Carlo based simulations of an expected outcome x from an experiment based on a theoretical prejudice (i.e. a null hypothesis), and the result of a measurement from data. The resulting distribution of the ensemble of Monte Carlo simulations is shown in Figure 8.1, along with an arrow that indicates the measurement result. On further inspection of the simulation, we find that 4.89% of the simulated experiments have a value of x that is greater than that indicated by the arrow. So the estimate of the probability of obtaining a result greater than the one observed in data, assuming that the Monte Carlo simulation is a valid representation of the measurement is: $\alpha = 4.89\%$. Note that there is a computational uncertainty in the probability estimate that is related to the number of simulated experiments N, and this uncertainty may be reduced by increasing N. With regard to the data in Figure 8.1, do we conclude that the data disagree with the theory, or do we conclude that the data are compatible? This brings us back to consider what is an acceptable level of false negatives we are willing to tolerate. If we wanted to ensure that the level of false negatives was 5%, then we would like to ensure that 95% of the time we would correctly identify an effect. As $\alpha < 5\%$, we would conclude in this case that the measurement was not compatible with an effect, and report a result inconsistent with H_0. Thus if H_0 was actually correct we would have been *unlucky* and reported a false negative. The chance of this happening would be about one in twenty. If on the other hand we are prepared to accept only a smaller rate of false negatives, say 1%, then we would conclude that this measurement is compatible

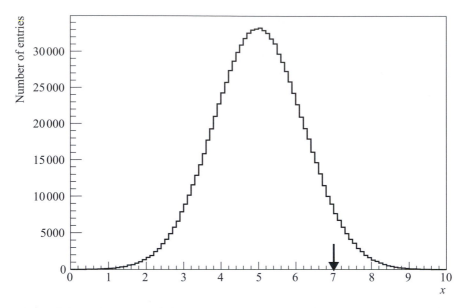

Figure 8.1 A histogram of simulated data compared with (arrow) the result of a measurement for the example discussed in the text.

with expectations of the simulation, and thus compatible with H_0. The conclusion drawn from the data will depend on the CL used for the test. It is important to set the CL, and hence maximum value of α prior to making the measurement to avoid biasing your conclusions according to any underlying prejudice you might have. To retrospectively decide upon on CL opens one up to experimenter bias, which is discussed in Section 6.6.

One may consider how serious a false negative is, and this depends on the situation. In the example of an experiment, a false negative will usually be re-tested at a later point in time, and a result will be refuted (unless the experimenter is biased by the prior existence of the false negative result). While it would be desirable to never obtain a false negative when testing for a particular medical condition, unfortunately that is an unrealistic goal. Thus in a medical situation, as a false negative may result in a patient with a serious condition being incorrectly given a clean bill of health, and suffering greatly as a result, one would like to minimise the rate of such an outcome. The balancing situation is discussed in the following section.

At this point it is worth introducing the concept of a ***p-value***. The p-value is the probability of obtaining a fluctuation in data under the null hypothesis that is as, or more, extreme than that observed by a given experiment. If the p-value is small, then one might consider rejecting the null hypothesis. This situation will arise if the result obtained in data is found to be in the extreme tail of possible outcomes under the null hypothesis. For example, if one were to find a *p*-value of ≤ 0.0027 for an observable with a Gaussian distribution, then this corresponds to a two-tailed fluctuation at the level of 3σ or more. This is not indicative of supporting the null hypothesis and an appropriate conclusion of such a result would be to reject H_0. The *p*-value in the previous example was 0.0489.

8.3 Testing if the hypothesis disagrees with data

In the previous section we discussed how to approach the question of testing if a hypothesis agrees with data, and discovered that one necessary condition was to determine an acceptable level or probability of assigning a false negative, i.e. of rejecting a measurement that was compatible with the existence of an effect. Now we turn to the reverse situation, that of disproving a theory. In this case the null hypothesis H_0 is that some effect that we are searching for exists, where in reality it does not. If we correctly identify that the hypothesis is invalid, then we will have obtained a true negative test. However, if we incorrectly establish the validity of H_0, then we will have obtained a *false positive* which is a type II error. As with the case of false negatives, the severity of a false positive depends on the scenario encountered.

8.4 Hypothesis comparison

Suppose we have two theories H_0 and H_1 proposed as descriptions of some data. We can compute $P(data|H_0)$ and $P(data|H_1)$ as defined by the theories, and we are able to specify a prior for both theories, $P(H_0)$ and $P(H_1)$. In general if H_0 and H_1 do not form the complete set of possible theories, then $P(data)$ is something that is not calculable. However, we can note that (see Eq. 3.5)

$$P(H_0|data) \propto P(data|H_0)P(H_0), \tag{8.1}$$

$$P(H_1|data) \propto P(data|H_1)P(H_1). \tag{8.2}$$

So while we can't compute $P(H_0|data)$ or $P(H_1|data)$, we can compute the ratio R between these posterior probabilities

$$R = \frac{P(H_0|data)}{P(H_1|data)} = \frac{P(data|H_0)P(H_0)}{P(data|H_1)P(H_1)}. \tag{8.3}$$

The relative probability R of theory H_0 to H_1 can be used to compare the two theories, and one may be able to determine which theory is a better description of the data. For example:

- if $R > 1$ then theory H_0 is preferred;
- if $R < 1$ then theory H_1 is preferred;
- if $R \simeq 1$ then there are insufficient data to discriminate between the two theories.

Example. Given a set of data resulting from the measurement of some observable

$$\Omega = \{-1.0, -0.9, -0.7, -0.1, 0.0, 0.1, 0.2, 0.5, 0.6, 1.0\}, \qquad (8.4)$$

where the total number of data $N = 10$, determine which of the following models is a better description of the data:

- H_0: the data are distributed according to a Gaussian PDF with a mean of zero and a width of one;
- H_1: the data are uniformly distributed.

Given this information, for each element of Ω we are able to compute R where for the sake of illustration we assume uniform priors $P(H_0)$ and $P(H_1)$, while

$$P(data|H_0) = G(\omega_i; 0, 1), \qquad (8.5)$$

$$P(data|H_1) = 1/N. \qquad (8.6)$$

Equation (8.3) can be used to compute R_i for a given element of data ω_i, and we are interested in the comparison of the total probability for the data, which is given by the product of the R_i,

$$R = \prod_{i=1}^{N} R_i, \qquad (8.7)$$

$$= \prod_{i=1}^{N} \frac{P(data|H_0)P(H_0)}{P(data|H_1)P(H_1)}, \qquad (8.8)$$

$$= \prod_{i=1}^{N} N G(\omega_i; 0, 1). \qquad (8.9)$$

Table 8.1 shows the results of each step of this calculation, resulting in the final value of $R = 140\,290$. As $R > 1$, theory H_0 is the preferred description of the data. In this example the prior dependence in Eq. (8.3) cancels. Generally the priors would not cancel if the form of $P(H_0)$ was chosen to be different from that of $P(H_1)$.

Consider now a second data sample

$$\Omega' = \{-4, -3, -2, -1, 0, 1, 2, 3, 4, 5\}. \qquad (8.10)$$

Table 8.1 *The values of P(data|theory) used to compute R_i for the sample of data Ω.*

ω_i	$P(data|H_0)$	$P(data|H_1)$	R_i
−1.0	0.242	0.1	2.42
−0.9	0.266	0.1	2.66
−0.7	0.312	0.1	3.12
−0.1	0.397	0.1	3.97
0.0	0.399	0.1	3.99
0.1	0.397	0.1	3.97
0.2	0.391	0.1	3.91
0.5	0.352	0.1	3.52
0.6	0.333	0.1	3.33
1.0	0.242	0.1	2.42

Just as before we are able to choose a prior for each of the models (for example flat or Gaussian), and compute contributions R_i to each of the events in Ω'. The value of the ratio R in this case is 3.6×10^{-13}. Thus for this data sample we find that $R \ll 1$, and hence the preferred theoretical description of the data is H_1, i.e. the data are flat.

8.5 Testing the compatibility of results

Often, instead of being faced with a theory and a measurement, we may encounter a situation where we have two measurements of some observable and we want to understand if they are compatible or not. If two measurements of the same observable are compatible, then any differences would be the result of statistical fluctuations in the data. If one or both of the measurements is affected by systematic effects, then the corresponding uncertainty σ_m would be the sum (usually in quadrature[1]) of statistical and systematic uncertainties. If we assume uncorrelated uncertainties between two measurements $m_1 = x_1 \pm \sigma_1$, and $m_2 = x_2 \pm \sigma_2$, then the difference between the measurements is given by

$$\Delta m = x_1 - x_2 \pm \sqrt{\sigma_1^2 + \sigma_2^2}, \tag{8.11}$$

$$= \Delta x \pm \sigma_m. \tag{8.12}$$

In order to test the compatibility of these measurements we must compare the magnitude of Δx to that of σ_m.

---

[1] By doing this one is making the tacit assumption that the systematic uncertainty is Gaussian in nature.

There is an intrinsic possibility that two correct measurements will be in dis-agreement. The larger the disagreement, the smaller the probability that the two results are compatible with each other. If we think in terms of Gaussian uncertain-ties, then it follows that by requiring two measurements to be within $\pm 1\sigma$ (68.3% CL) in order to consider them to be in agreement, we would classify 31.7% of legitimate results as being as incorrect or inconsistent. A more sensible way of looking at this problem is to define an acceptable level of mis-classification, given by $1 - CL$, and use this to test if the results are compatible. Typically we consider two results to be incompatible if they differ by more than $3\sigma_m$ from each other. Given a large number of compatible measurement comparisons we would expect this criteria to give an incorrect classification of agreement about three in every 1000 times. In general if we find results that are between 2 and 3σ apart, we tend to scrutinise them a little more closely to understand if there are systematic effects that may have been overlooked in one or both of the measurements. If both results hold up to scrutiny, then we conclude that they must be compatible.

It is completely natural to expect that about three in every 1000 measurements made of some observable will be more than 3σ from the true value of that observ-able. The consequence of this fact is that for about every 300 correct measurements you make, one of them can be expected to be statistically inconsistent with pre-vious or subsequent measurements at a level of 3σ. This does not mean that that measurement is wrong, this is simply a reflection of the syntax used to describe a measurement.

Example. If some observable O has a true value of μ, and it is possible to perform a measurement of that observable which would result in an expected standard deviation of σ, where systematic effects on the measurement are negligible, then we can perform an ensemble of measurements and observe how these results compare to the expectation of $\mathcal{O} = \mu$. If we perform 1000 measurements, then according to Table 6.1, we would expect on average to obtain 2.7 measurements that are more than 3σ from μ. Figure 8.2 shows the results of 1000 such measurements where $\mu = 0$ and $\sigma = 1$. We can see in this instance that there are two measurements that are between 3σ and 4σ from μ, a result consistent with expectations.

8.6 Establishing evidence for, or observing a new effect

In general there are no globally accepted confidence levels that are used to establish the existence of a new effect. If we analyse data, with a null hypothesis that no effect exists, and find that the data are incompatible with that hypothesis at some confidence level, then we can interpret that result in a reasonable way. The reader should keep in mind that different fields use different confidence levels to test the

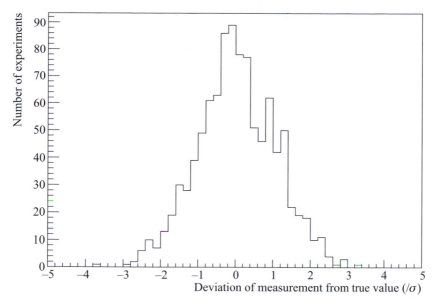

Figure 8.2 The resulting distribution on performing 1000 measurements of an observable $\mathcal{O}$.

significance of a result, and so 'standard practice' for one specialisation is probably not directly applicable to another.

For example, in particle physics, one often chooses to consider a deviation from the absence of an effect at the level of 3σ as '*evidence*' for the existence of that effect. Assuming that the measurement uncertainties are Gaussian in nature, this translates into a claim of evidence for some effect to be a false positive statement about three in every 1000 correct measurements. One may require a more stringent classification for the existence of an effect than at the level of 3σ. If one finds a deviation from the absence of an effect at the level of 5σ, then we might regard this as '*observation*' of an effect. Only once every million measurements will we report a false positive result at this level. In contrast another field may choose to use a 2σ threshold to establish an effect. If you insist on describing a result with a phrase such as evidence for or observation of then you should remember that this label has little meaning, and the interpretation of the phrase is ambiguous. The proper way to address this is to specify the result in terms of the CL or p-value used in order to draw your conclusions.

8.6.1 Peak finding

Broadly speaking there are two types of peak finding problems: (i) searching for a peak at some known position in parameter space, and (ii) searching for a peak at some unknown location in parameter space (see Section 8.6.2). An example of the former scenario is the decay of an unstable particle into some identified final state, where that state is reconstructed from measurements of energy deposits in one or more detectors. For example, a π^0 meson can decay into two electrons and a photon.[2] If one has accurate enough measurements of the energy and three-momenta of each of the decay products then one can compute the invariant mass of the π^0 from data. The mass of the π^0 is known to be approximately 135 MeV/c^2 (Beringer *et al.*, 2012), and the width of any peak in the reconstructed invariant mass distribution is given by detector resolution. Given this information it is possible to search through data and compute the invariant mass m_0 for all combinations of an $e^+e^-\gamma$ final state. This can be obtained using the relativistic energy–mass relation $E^2 = m_0^2 c^4 + p^2 c^2$, where c is the speed of light in vacuum, E is the sum of particle energies and p is the sum of particle three-momenta. If there are $\pi^0 \to e^+e^-\gamma$ events in the data, then this should be evident by looking at this histogram where one should see a peak at the appropriate location. The other combinations of events that enter the histogram would be from random combinations of candidate electrons, positrons and photons. These background events would have a flat distribution. In order to ascertain if one had a signal or not, one would have to determine if a peak at the appropriate mass was large enough to be unlikely to correspond to a fluctuation of the background. Hence the null hypothesis in this scenario would be that the histogram only contains background events, and there is no signal. An example of this problem can be found in Figure 8.3, where there is a peak visible at 135 MeV/c^2 as expected. The width of this distribution is the result of the finite resolution of the detector. One can see the signal is significant without having to resort to a significance or p-value calculation.

8.6.2 Trial factors and nuisance parameters

An example of the second type of search can be seen by once again considering Figure 8.3. The π^0 peak at 135 MeV/c^2 was expected and we can ignore that sample of events for the following discussion. Having done that one can see that there are two clusters of events which appear near 140 and 160 MeV/c^2. Are these also particles? In this simulation there are 100 background and 100 $\pi^0 \to e^+e^-\gamma$ signal events generated, and the histogram has 60 bins. Hence on average each bin has an expectation of 5/3 background events in it, and any excess beyond that could be considered a potential signal. Five background events are expected for both of the excesses, where each excess spans three adjacent bins. There are 12 events appearing in each cluster bins. One can ask what the probability is for the

---

[2] This particular decay, $\pi^0 \to e^+e^-\gamma$, is sometimes referred to by name as a π^0 Dalitz decay.

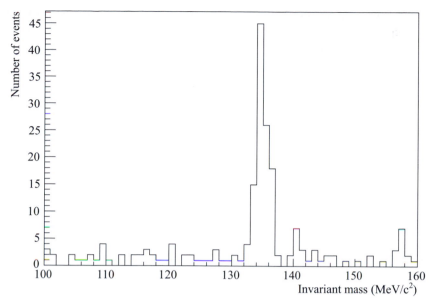

Figure 8.3 The distribution of events obtained from a simulated experiment illustrating the peak finding problem for $\pi^0 \rightarrow e^+e^-\gamma$ decays.

background level to fluctuate up to the observed level of events. The probability of observing 12 or more events given $\lambda = 5$ for a Poisson distribution is 0.55%. This *p*-value is quite small given the *CL* values often used to place bounds on unseen effects, and one might be tempted to report that another particle (or two) exists in the data. The problem is that while only one sample of data is being analysed, we are looking for peaks in many different sets of bins and we need to take into account that we are performing a number of simultaneous searches when we compute the *p*-value for any apparent excess. In general one can compute a trial factor, the ratio of the probabilities of observing an excess at a given point and observing an excess anywhere in a range, to help interpret such a situation. In order to determine the trial factor and hence compute the correct *p*-value for a given result, one can perform an ensemble of background only simulations to determine the largest upward fluctuations of background, and compare that with data. This can be both time and resource intensive. It is also possible to approach the problem from a more formal approach as described by Gross and Vitells (2010) and references therein. This class of problem is referred to as 'hypothesis testing when a nuisance parameter is present only under the alternative', or colloquially as the *look elsewhere effect*. The term **nuisance parameter** refers to a parameter that is not the main result, but must be considered in the extraction of the main result from data, for example see Edwards (1992). In the example discussed above, the main parameter of interest when searching for a peak is the mass, and the width of the mass peak would be classed as a nuisance parameter. When computing a Bayesian constraint any nuisance parameters can be removed from the problem by integrating them out. This process is sometimes referred to as marginalisation, and the resulting probability distribution is called the marginal distribution.

8.7 Case studies

This section discusses a few case studies that can be understood by applying some of the techniques discussed in this chapter.

8.7.1 Clinical test result

A test for an infection returns a positive result 99% of the time for someone with an infection. The same test reports a positive result 1% of the time for patients who are not infected. If 0.1% of the population are infected, what is the probability that someone with a positive test result is actually infected?

From this information we can use Bayes' theorem to compute the probability that someone with a positive test result is actually infected. Firstly we note that the prior probabilities for the hypotheses *infected* and *healthy* are given by

$$P(infected) = 0.001, \tag{8.13}$$

$$P(healthy) = 0.999. \tag{8.14}$$

We also note that

$$P(positive\ test\ result|infected) = 0.99, \tag{8.15}$$

$$P(positive\ test\ result|healthy) = 0.01. \tag{8.16}$$

The posterior probability in this instance is given by

$$P(infected|positive\ test\ result) = \frac{0.001 \times 0.99}{0.001 \times 0.99 + 0.999 \times 0.01}, \tag{8.17}$$

$$= 0.09. \tag{8.18}$$

Hence the probability that someone with a positive test result is infected is only 9%. This would indicate that the test in question has a large false positive rate and is not particularly useful. If one assumed that a false positive was a random effect, but that an infected patient would always give a positive result, then it could be possible to re-test a patient several times to reduce the level of false positives. However, with such a poor performance, re-testing would not be desirable.

8.7.2 The boat race

Before the 2010 Oxford–Cambridge boat race, there had been 70 winning crews taking the Middlesex side, and 75 winning crews rowing on the Surrey side. Having won the toss, knowing these data, what side of the river should the Oxford crew decide to race on? The race commentator proposed that one would *naturally* choose the Surrey side as that *obviously* would give the crew an edge. Is this assertion well founded given the available data?

The total number of events is 145. Lets assume that the Surrey side is the winning side, where the probability for winning is given by $\epsilon_{Surrey} = 75/145 = 51.7\%$. The binomial error on this probability, see Eq. (6.33), gives $\sigma_{Surrey} = \sqrt{\epsilon_{Surrey}(1 - \epsilon_{Surrey})/145} = 4.1\%$. So we would expect the Surrey side to produce a winner in $(51.7 \pm 4.1)\%$ of the races, and consequently we would expect the Middlesex side to produce a winner in $(48.3 \pm 4.1)\%$ of races. Given the uncertainties on these probabilities we must simply draw the conclusion that the Oxford crew should do whatever they want as there is no evidence that starting from either side gives a statistically significant advantage.

8.8 Summary

The main points introduced in this chapter are listed below.

(1) There is an element of subjectivity required in order to test any hypothesis; however, at the same time it is important to make such a test objectively and to report the outcome in such a way that another scientist can discern the outcome of your efforts. The first step is to clearly define a null hypothesis, and decide how this will be tested.

(2) If we conclude that an effect is real, then it is most likely that subsequent measurements will be made, and in turn these more precise measurements could confirm our earlier result, or refute any such result as a statistical fluctuation or as an incorrect measurement. For this reason we would like to minimise our possibility of making an incorrect judgment (in this case creating a false positive).

(3) If, on the other hand, we search for an effect and conclude it does not exist, then there will be less motivation for one to repeat a measurement. Again we do not want to set our standards so high that we will ignore a real effect should we measure it (in this case creating a false negative). For these reasons, we must strike a careful balance between these two competing requirements of minimising a false positive statement, and minimising a false negative statement.

(4) When testing a hypothesis, one should make sure that before making a measurement we

- formulate the hypothesis precisely, and thus define the alternate hypothesis,
- identify an acceptable mis-classification rate for the test, for example set a confidence level of 90%, 95%, or 99%, etc., to reject compatibility of the measurement with some null hypothesis,
- make the measurement,
- compare the measurement with the null hypothesis,
- clearly report what has been determined from this measurement, stating the criteria used to reach that conclusion.

(5) It is possible to compare two results and determine if they are compatible. In general if the two results lie within 3σ (1σ) of each other they are considered (very) compatible.

(6) There are no defined limits for establishing if an effect is real or not: however, working hypotheses that are often used are as follows. There is evidence for an effect if it is possible to establish from data, that the hypothesis of no effect is inconsistent at the level of 3σ. A stronger statement can be made which is sometimes referred to as observation. In order to claim an observation, one often requires a 5σ deviation from the hypothesis of no effect existing. Any result quoted should have the corresponding CL or p-value stated in order to allow a reader to understand the meaning of any descriptor used. It is the CL or p-value of a result that is the important piece of information.

Exercises

8.1 Given the data $\Omega(x) = \{0.00, 0.10, 0.15, 0.20, 0.21\}$, use Bayes' theorem to compare the hypothesis that the data are uniformly distributed against the hypothesis that the data are distributed with a mean of 0.15 and a Gaussian spread of 0.9. Which hypothesis agrees better with the data?

8.2 A test for an infection returns a positive result 98% of the time for someone with an infection. The same test reports a positive result 0.05% of the time for patients who are not infected. If 0.01% of the population are infected what is the probability that someone with a positive test result is actually infected? Is this a good test?

8.3 A test for an infection returns a positive result 99.99% of the time for someone with an infection. The same test reports a positive result 0.01% of the time for patients who are not infected. If 0.1% of the population are infected what is the probability that someone with a positive test result is actually infected? Is this a good test?

8.4 A test for an infection returns a positive result 99.99% of the time for someone with an infection. The same test reports a positive result one in a million times for patients who are not infected. If 10% of the population are infected what is the probability that someone with a positive test result is actually infected? Is this a good test?

8.5 The following results are all measurements of the height of the same person: 1.83 ± 0.01 m, 1.85 ± 0.01 m, 1.87 ± 0.01 m. Are they compatible with each other?

8.6 The data given by $\Omega = \{0, 1, 2, 4, 6\}$ is for a Poisson process. Which value of λ, either 3 or 4, is a better description of the data?

8.7 The data given by $\Omega = \{1, 2, 3, 5, 7\}$ is for a Poisson process. Which value of λ, either 3 or 4, is a better description of the data?

8.8 Which hypothesis agrees with the data $\Omega = \{1, 2, 3, 4, 5\}$ better, a binomial distribution with $p = 0.4$, or a Gaussian distribution with a mean of 3 and width of 1?

8.9 What is the p-value obtained for a rare decay experiment where one expects one event and observes five? Does this result contradict expectations?

8.10 What is the p-value obtained for a rare decay experiment where one expects one signal event, with two background and observes five events? Does this result contradict expectations?

8.11 When searching for an effect expected to be small, at the level of five events in your data, you observe 15 events. What is the p-value for this outcome? Does this result contradict expectations?

8.12 What is the p-value obtained for a Gaussian observable to be non-zero given the measurement $x = 0.5 \pm 0.1$?

8.13 Measurements of the W boson mass from the MARK-2 and OPAL experiments are 91.14 ± 0.12 and 91.1852 ± 0.0021 GeV/c^2, respectively. At what level do they agree with each other?

8.14 Measurements of the mass of the new particle discovered in summer 2012 by the ATLAS and CMS experiments at CERN are $126 \pm 0.4 \pm 0.4$ and $125.3 \pm 0.4 \pm 0.5$ GeV/c^2, respectively. The first uncertainty quoted is statistical and the second is systematic in nature. At what level do these mass measurements agree with each other?

8.15 Can both of the following results be correct (explain your reasoning) $x < 1.0 \times 10^{-6}$ at 90% CL, and $x = (1.5 \pm 1.0) \times 10^{-6}$, where $x \geq 0$?

9

Fitting

9.1 Optimisation

The process of ***fitting*** a sample of data $D(\underline{x})$ containing N entries involves iterative comparison of data and a theoretical model $\mathcal{P}(\underline{x}, \underline{p})$ assumed to describe those data.[1] The $\underline{x}$ are discriminating variables to be used in the fit, taking the value $\underline{x}_i$ for the ith event, and $\underline{p}$ are the parameters of the theoretical model. The parameters are allowed to vary in the fit to data, i.e. each iteration of the optimisation process uses a different value for $\underline{p}$, in order improve the compatibility of data and model. This process requires the definition of a test statistic T that is used to quantify how well the model and data agree and to then vary $\underline{p}$ in such a way as to try and obtain an improved model description of the data. The test statistics described in this section are χ^2 and likelihood based. In both cases we numerically minimise the test statistic summed over the data. So in order to perform a fit to data we perform an optimisation process in the parameter space $\underline{p}$ which involves minimising the sum

$$S = \sum_{i=1}^{N} T[D(\underline{x}_i), \mathcal{P}(\underline{x}, \underline{p})]. \tag{9.1}$$

In order to converge on a solution one has to start with an initial estimate of the parameter set to evaluate S. Having done this one then determines a new estimate of $\underline{p}$ following a pre-defined rule, at each step evaluating the corresponding S. After making an initial set of estimates of the parameter set, it is normally possible for the algorithm to determine in which direction a more optimal set lies. Having done this, the algorithm will perform another search starting from a point $\underline{p}'$ that is closer to the assumed minimum than the previous one. This process is repeated

---

[1] Depending on the method we use, we are able to fit the data on an event-by-event basis, or by binning the data in finite intervals.

until such time as the optimisation algorithm has a sufficiently small ***step size*** given by

$$\delta = |\underline{p} - \underline{p}'|. \tag{9.2}$$

When δ is smaller than some minimum step size or distance ϵ in the parameter space, the optimisation is said to have converged on a minimum value $\underline{p}_{\text{min}}$. Having found a minimum $\underline{p}_{\text{min}}$ the final step is for the algorithm to determine the corresponding uncertainty $\delta\underline{p}_{\text{min}}$, which depends on the test statistic that is being minimised.

General issues with numerical optimisation procedures are that they do not always distinguish between local and global minima, and they do not always converge to a minimum. If a minimum is found, such that $\delta < \eta$ where η is a convergence requirement, the algorithm considers this point in space to be the minimum. Further tests must be made in order to determine if one has found a local or global minimum. The procedure to validate that a minimum is global involves scanning the start parameters for the fit, and repeating the minimisation. It is not always practical to perform such a parameter scan; however, there are certain circumstances where it is absolutely necessary, for example fits with large numbers of parameters where there are large correlations between them.

One thing to be borne in mind is that the number of computational steps required in order to converge to a local or global minimum will depend on how close the initial values of $\underline{p}$ are to the values corresponding to the minimum and on the step size used to iterate $\underline{p}$. It is possible that a fit may not converge to a minimum if the values of $\underline{p}$ are far from the values at a minimum. There are many ways to minimise a quantity, each have their own pitfalls, and only two examples are described in the following to illustrate the process of optimisation. In practice one normally uses a more sophisticated approach to determine the optimal set (or fitted) parameters $\underline{p}$.

9.1.1 Gradient descent method

Consider an m-dimensional parameter set that is to be fit to some model given by $\mathcal{P}$ for some data sample. The sum S used to compare the data with the model is given by Eq. (9.1). Starting from a point in the parameter space $\underline{p}$, one can numerically compute an estimate for the gradient

$$g(\underline{p}) = \frac{\partial S}{\partial \underline{p}}, \tag{9.3}$$

from one (or more) neighbouring point(s) $\underline{p} + \Delta \underline{p}$. In practice we can only determine an approximation of the gradient

$$g(\underline{p}) \simeq \frac{\Delta S}{\Delta \underline{p}}. \tag{9.4}$$

Having determined the gradient at the point $\underline{p}_j$, one can use this to estimate a new point $\underline{p}_{j+1}$ some small distance $\underline{\epsilon}$ from $\underline{p}_j$ (the sign of ϵ_j depends on the sign of the gradient)

$$\underline{p}_{j+1} = \underline{p}_j + \underline{\epsilon} \cdot g(\underline{p}). \tag{9.5}$$

In general, one expects $\underline{p}_{j+1}$ to be a better estimate of the true minimum than $\underline{p}_j$. One can continue iterating on this process, each time estimating a new parameter set until such time as the estimated gradient is close enough to zero to be considered the minimum. In general the condition for having determined the minimum is

$$\frac{\Delta S}{\Delta \underline{p}} = 0, \tag{9.6}$$

and the value of S either side of the stationary point found in the parameter space can be used to distinguish between possible maxima, minima, and points of inflection. As this search is numerical, sometimes the new estimate $\underline{p}_{j+1}$ can have a value that is further from the minimum than $\underline{p}_j$. This can occur as the step size $\underline{\epsilon}$ is a pre-determined parameter for the search and may simply be too large. If subsequent iterations fail to re-converge on a minimum, this optimisation procedure can fail to converge. For complicated models, failure to converge may not be uncommon, and as such one should take care to choose a reasonable starting parameter set $\underline{p}_0$.

9.1.2 Parameter scan

In certain situations it may not be possible to optimise or fit the values of all parameters that a model depends upon, or one may want to understand the change in S as a function of a particular parameter without resorting to an optimisation algorithm. In such scenarios one can perform a parameter scan. This involves stepping through the values of a parameter of interest p from some minimum value through to some maximum value. The minimum and maximum are chosen such that the best fit value that one is attempting to determine is bound by these limits. At each point between p_{min} and p_{max} the sum S is computed. Hence one can plot S as a function of p in order to determine the optimal value of the parameter via such a *scan*. If the model depends on more than one parameter then the ancillary parameters should be optimised at each value of p used in the scan. An example of this approach is given in Section 9.2.1.

Table 9.1 *Individual measurements of a parameter S as described in the text. The uncertainties* σ_S *considered are statistical only.*

S	0.662	0.625	0.897	0.614	0.925	0.694	0.601
σ_S	0.039	0.091	0.100	0.160	0.160	0.061	0.239

9.2 The least squares or χ^2 fit

In order to perform a χ^2 fit, one typically bins data such that there are at least five to ten events that contribute to each bin. As a result the data are generally not binned in samples that are of equal size in the discriminating variable space. The test statistic used for this type of fit is a χ^2 constructed between the data D_i and theoretical model describing the data $\mathcal{P}_i$, i.e.

$$\chi^2 = \sum_{i=1}^{n} \left(\frac{D_i - \mathcal{P}_i}{\sigma(D_i)} \right)^2, \qquad (9.7)$$

(c.f. Eq. (5.44)) where the sum is over the number of bins. The parameter ($\underline{p}$) and discriminating variable ($\underline{x}$) dependence have been suppressed here, and are implied. The quantity $\sigma(D_i)$ is the uncertainty on the datum D_i, which in the case of an event yield is given by the corresponding Poisson error on that yield (or Gaussian in the limit of large statistics). As by definition the χ^2 distribution is normalised by the uncertainty from data, the 1σ error on a result is given when the χ^2 changes by one unit from the minimum value χ^2_{min}. Once you have a data sample to fit, the only remaining issue is the choice of model $\mathcal{P}$. Some commonly used PDFs are discussed in Appendix B, in addition to those already encountered earlier in the book. The test statistic in Eq. (9.7) is sometimes referred to as a least squares statistic. In this book the terminology χ^2 fit is generally used in the context of an arbitrary model $\mathcal{P}$ describing the data, where a numerical minimisation is performed to obtain the optimal set of fit parameters p, and least-squares optimisation is used in the context of problems that are solved by analytic means (see Section 9.3). This distinction is artificial and introduced here to distinguish between the two ways of solving a problem using this type of test statistic.

9.2.1 Example: determining the average of a set of measurements

Consider the situation where one has several measurements of the value of some quantity S as shown in Table 9.1. This quantity is the measure of difference between matter and antimatter decaying from an initial state called a B meson into a final state involving so-called charmonium particles and a strange particle (a kaon). The data are taken from a journal article written by collaborators working on a high-energy physics experiment called BABAR (Aubert *et al.*, 2009). The parameter S has to be zero for matter and antimatter to behave in the same way for these measurements.

Based on this information, *what is the average value of S?* While it is possible to compute a weighted average using the formalism outlined in Section 6.4.1, it is also valid to consider using a fit or scan to determine the average value of a set of measurements. The advantage of using a scan-based approach to compute the average is the retention of more information concerning the observable we are trying to measure. Instead of a single number to represent the uncertainty, one has a curve that can be used to determine confidence levels of arbitrary significance.

Each column in the table corresponds to an event i that contributes to a χ^2 sum. As we are trying to determine the most optimal value of S, the model is given by the assumed value of that parameter, in other words we perform a parameter scan of S, from S_{min} to S_{max}, and for each point in this range we can compute the χ^2 sum

$$\chi^2 = \sum_{i=1}^{7} \left(\frac{S_i - S}{\sigma_{S_i}} \right)^2, \tag{9.8}$$

where S_i and σ_{S_i} are the central values and uncertainties for the ith event. This is shown in Figure 9.1. The minimum value of the χ^2 is at $S = 0.690$, and the corresponding error (how far one moves away from the minimum value of S in order to obtain a change in χ^2 of one from the minimum value) is 0.028; thus the average value of the data in the table using this method is $S = 0.690 \pm 0.028$. For comparison, the average value obtained in the original reference is $S = 0.687 \pm 0.028$. Note that the method used for the average computed here is not as sophisticated as that in Aubert *et al.* (2009), which explains the small difference obtained between these two results. The results of the method outlined here and the one used in the original reference give essentially the same average given the precision of this set of measurements.

It turns out that the parameter S is a function of a more fundamental quantity, an angle β. Given that $S = \sin(2\beta)$ one can repeat the parameter scan in terms of the angle β by replacing S in Eq. (9.8) by $\sin(2\beta)$. For each assumed value of β in $[0, 360]°$ one can compute $\sin(2\beta)$ and hence determine the χ^2 sum as before. The corresponding average value obtained for β in the first quadrant is $(21.8 \pm 1.1)°$.

If instead of performing a scan in S or β, we chose to use a minimisation algorithm, we would have arrived at the same results as obtained from Figure 9.1. However, in performing just a fit to the data, we would have obtained only a central value, and uncertainty estimate for the average value of the observable. By performing both the fit, and the parameter scan we have an estimate of the optimal value of the observable, and a visual representation of the behaviour of the test

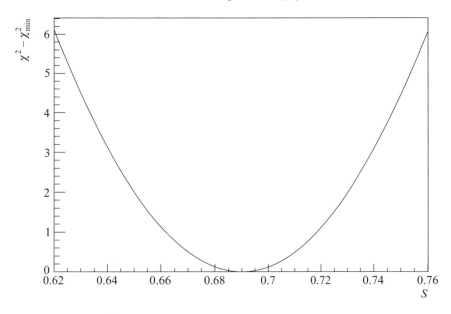

Figure 9.1 The χ^2 parameter scan in terms of S for the data given in Table 9.1.

statistic in the vicinity of this value. This enables us to verify that the test statistic is smoothly varying (i.e. well behaved) in the vicinity of the minimum.

Having obtained the optimal result from a minimisation process, it is possible to compute the probability of obtaining that result given the number of degrees of freedom, using $P(\chi^2, \nu)$ from Eq. (5.43). This is an indicator of how well the model agrees with data and is often referred to as the goodness of fit, or GOF (see Section 5.5), and it is something that should be considered when validating a result. For example in the case described here we have obtained the average result $S = 0.690 \pm 0.028$. The sum of χ^2 deviations from the average for this data (often abbreviated as the χ^2) is $\chi^2 = 7.83$. There are $n - 1$ degrees of freedom, as there are $n = 7$ data used in the evaluation of the χ^2, and the total sum is constrained by the number of data. Thus there are six degrees of freedom ($\nu = 6$). The χ^2 probability for this situation is given by $P(\chi^2, \nu) = 0.25$, which means that the result has a reasonable outcome.

If we had obtained a result where $\chi^2 \sim 0$, then the uncertainties on each of the individual data would have been overestimated. This would indicate that one has probably inflated or overestimated the errors in order to improve the level of agreement of the individual measurements, or alternatively that the data are highly correlated, and can not be combined without accounting for the correlation. Similarly had we obtained $\chi^2/\nu \gg 1$, then we would conclude that the data are not in good agreement, as again one would find that $P(\chi^2, \nu) \simeq 0$. In either of these extremes we have to try to understand if it is meaningful to combine the results (this is best done before attempting any such combination where possible). If there is a particular piece (or several pieces) of data that dominates the χ^2 then studying that input in more detail would be in order. If on inspection it turned out that the data for a given event are suspect, for example the measurement was wrong, then that data point may be ignored. If, however, that experiment appeared to be reasonable, then it is not appropriate to discard the data as the result may simply be a statistical outlier.

9.3 Linear least-squares fit

The procedure adopted in Section 9.2 can be used to study general situations where the theoretical model described by $\mathcal{P}$ is arbitrary as illustrated through the previous example of determining average values of S and β. Often one can take an analytical approach to solve a given problem. One situation that often arises is the case of comparing data with the model $\mathcal{P}_i = ax_i + b$, where the uncertainty on each point $\sigma(y_i)$ is some constant value denoted by σ. In this case we can write Eq. (9.7) as

$$\chi^2 = \sum_{i=1}^{n} \left(\frac{y_i - ax_i - b}{\sigma(y_i)} \right)^2 = \frac{1}{\sigma^2} \sum_{i=1}^{n} (y_i - ax_i - b)^2 , \qquad (9.9)$$

where we have replaced D_i by y_i and made the appropriate substitution for $\mathcal{P}_i$. The task at hand now is to minimise χ^2 with respect to both a and b in order to determine the optimal values of the slope and offset of our model. In order to do this we differentiate χ^2 with respect to a and b, and simultaneously solve for values that correspond to the point where both derivatives are zero.[2] In the simplified case where the uncertainties on the y_i are all some constant value, we can remove the constant $1/\sigma^2$ from the problem. If we consider the derivative with respect to a first, this is just

$$\sigma^2 \frac{\partial \chi^2}{\partial a} = \sum_{i=1}^{n} \frac{\partial}{\partial a} (y_i - ax_i - b)^2 , \qquad (9.10)$$

$$= \sum_{i=1}^{n} -2x_i(y_i - ax_i - b), \qquad (9.11)$$

$$= -2 \sum_{i=1}^{n} x_i y_i - ax_i^2 - bx_i , \qquad (9.12)$$

$$= -2n(\overline{xy} - a\overline{x^2} - b\overline{x}). \qquad (9.13)$$

---

[2] In general, one can use the second derivative or numerical means to establish the nature of the turning point, and ensure one has located a minimum.

Similarly for the derivative of χ^2 with respect to b one obtains

$$\sigma^2 \frac{\partial \chi^2}{\partial b} = -2n(\bar{y} - a\bar{x} - b). \tag{9.14}$$

As the optimal solution exists for

$$\frac{\partial \chi^2}{\partial a} = 0, \text{ and } \frac{\partial \chi^2}{\partial b} = 0, \tag{9.15}$$

we need to simultaneously solve (ignoring constant multipliers)

$$\overline{xy} - a\overline{x^2} - b\bar{x} = 0, \tag{9.16}$$

$$\bar{y} - a\bar{x} - b = 0, \tag{9.17}$$

for a and b. The results of this are

$$a = \frac{\overline{xy} - b\bar{x}}{\overline{x^2}} = \frac{\overline{xy} - \bar{x}\,\bar{y}}{\overline{x^2} - \bar{x}^2}, \tag{9.18}$$

$$b = \bar{y} - a\bar{x}. \tag{9.19}$$

Hence, for the situation where we want to determine the coefficients of a straight line fit to data, where the data y_i have equal uncertainties, and the abscissa values x_i are precisely known, we can analytically solve for the slope and intercept parameters without having to perform a numerical optimisation.

Using the combination of errors procedure outlined in Chapter 6 on Eqs. (9.18) and (9.19) it can be shown, for example see Barlow (1989), that

$$\sigma^2(a) = \frac{\sigma^2}{N(\overline{x^2} - \bar{x}^2)} \tag{9.20}$$

$$\sigma^2(b) = \frac{\sigma^2 \overline{x^2}}{N(\overline{x^2} - \bar{x}^2)}. \tag{9.21}$$

In general the least squares method can be written in matrix form as

$$\chi^2 = \Delta^T V^{-1} \Delta, \tag{9.22}$$

where V is the covariance matrix and Δ is a column matrix of difference terms of the form $x_i - f(x_i, p)$, and p are parameters of the model f. From this general form it is possible to derive the weighted averaging procedure introduced in Chapter 6.

The use of least squares regression is discussed in more detail for example in the books by Barlow (1989); Cowan (1998); Davidson (2003); James (2007).

9.4 Maximum-likelihood fit

It is possible to define a likelihood function $\mathcal{L}$ that uses the PDFs $\mathcal{P}$ to describe the distribution of discriminating variables $\underline{x}$ in data. As the functions $\mathcal{P}$ are normalised so that the total probability of an event is unity, we can write the likelihood for an event as

$$\mathcal{L}_i = \mathcal{P}(\underline{x}_i). \tag{9.23}$$

The quantity $\mathcal{P}(\underline{x}_i)$ is the probability assigned by the model for the ith event. If the event $\underline{x}_i$ is certainly signal, then the part of $\mathcal{P}(\underline{x}_i)$ corresponding to signal will be one. Similarly if the event is definitely not signal then the part of $\mathcal{P}(\underline{x}_i)$ corresponding to signal will be zero. One can make similar statements for background, or indeed any other type or class of event described by the model $\mathcal{P}$. Usually the probability assigned to an event for a given component j is between these two extreme values, so in general $0 \le \mathcal{P}_j(\underline{x}_i) \le 1$. In order to express these possibilities mathematically we need to consider the situation when there are distinct components in the model. For example, if there are m components in a model, then the likelihood for an event i is given by a sum over these m components

$$\mathcal{L}_i = \sum_{j=1}^{m} f_j \mathcal{P}_j(\underline{x}_j), \tag{9.24}$$

where f_j are the fractions of the different components. For a single event the f_j are interpreted as the probability that an event is of type j. Thus in order to conserve probability we require[3]

$$\sum_{j=1}^{m} f_j = 1. \tag{9.25}$$

So far we have only considered a single event that may be one of m different possible types. In reality there is a limited amount of information that can be gleaned from a single event and we are usually faced with interpreting results from data samples of many events. As each event is independent, the likelihood for a data set containing N events is the product of the likelihoods for the individual

---

[3] It is possible to use the normalisation constraint of Eq. (9.25) to reduce the total number of parameters by one, resulting in the final component fraction $f_m = 1 - \sum_{i=1}^{m-1} f_j$.

events

$$\mathcal{L} = \prod_{i=1}^{N} \mathcal{L}_i,$$ (9.26)

$$= \prod_{i=1}^{N} \sum_{j=1}^{m} f_j \mathcal{P}_j(\underline{x}_i).$$ (9.27)

It can be troublesome to numerically compute $\mathcal{L}$ for large samples of data. The test statistic that is usually optimised is

$$-\ln \mathcal{L} = -\ln \prod_{i=1}^{N} \mathcal{L}_i,$$ (9.28)

$$= -\ln \prod_{i=1}^{N} \sum_{j=1}^{m} f_j \mathcal{P}_j(\underline{x}_i),$$ (9.29)

$$= -\sum_{i=1}^{N} \sum_{j=1}^{m} f_j \mathcal{P}_j(\underline{x}_i),$$ (9.30)

which is easier to compute numerically and follows the form of Eq. (9.1). All that remains is to define the fit model by choosing the $\mathcal{P}_j$ distributions. A number of PDFs that can be used to construct models are discussed in Chapter 5 and Appendix B.

9.4.1 Extended maximum-likelihood fits

If one is interested in determining event yields n_j of a set of categories, instead of fractions, then the Poisson nature of those yields needs to be taken into account. In such circumstances the likelihood function given in Eq. (9.27) is modified by a Poisson term that depends on the total fitted event yield $n' = \sum n_j$, where n_j is the number of fitted events of type j, to give (Barlow, 1990)

$$\mathcal{L} = \frac{e^{n'}}{N!} \prod_{i=1}^{N} \sum_{j=1}^{m} n_j \mathcal{P}_j(\underline{x}_i),$$ (9.31)

The total number of fitted events n' does not have to exactly match the total number of events N. The fit has the ability to converge on some minimum value close to, but not necessarily satisfying $n' = N$. More details on the usage of this form of likelihood fit can be found in the literature (Barlow, 1990; Cowan, 1998). If the event yields n_j are observables of primary interest then an extended maximum-likelihood fit provides a more convenient approach to use than the original form.

9.4.2 Interpreting the result of a likelihood fit

In general, as with the case of least squares, one can analytically solve for the optimal value of a parameter for a given problem by requiring

$$\frac{\partial \mathcal{L}}{\partial p} = 0, \tag{9.32}$$

and subsequently by using the second derivative or numerical means to ensure that the stationary point is a maximum. The maximum will provide the most likely value of p which is denoted here by p_0. In general the variance on p is given by the integral equation

$$(\Delta p)^2 = \frac{\int (p - p_0)^2 \mathcal{L} dp}{\int \mathcal{L} dp}, \tag{9.33}$$

which follows directly from Eq. (5.2).

In practice it is often cumbersome to analytically compute the optimal value and variance for a parameter for a complicated fit model. In such a scenario one must resort to numerical means of evaluating the best fit value. On computing the value of $-\ln \mathcal{L}$ for a data set with a given assumed parameter set we are able to compute a number that is related to the probability of the agreement of the data to the assumed model with the assumed parameters. After completing this initial step, one performs an optimisation of $-\ln \mathcal{L}$ as described in Section 9.1. As a natural part of the optimisation process the fit will converge on some optimal result denoted by $-\ln \mathcal{L}_0$, that corresponds to a minimum value of $-\ln \mathcal{L}$. This is the most probable result that our algorithm has been able to identify. This is a single point in parameter space, and tells us nothing about the uncertainty on that point in space in terms of the parameters. Furthermore, if there is more than one parameter in the parameter set, then one has to worry about how correlated those parameters might be with each other, and that in turn can complicate the determination of the uncertainty on a parameter. For such a scenario the problem has to be solved in a multi-dimensional space.

If we consider an observable p, then we expect the uncertainty on p to be distributed according to a Gaussian for large data samples. Hence, for large samples, we expect

$$\mathcal{L}(p, \mu, \sigma) = \frac{1}{\sigma \sqrt{2\pi}} e^{-(p - \mu)^2 / 2\sigma^2}, \tag{9.34}$$

with some mean value μ and standard deviation σ providing an estimate of the parameter p and its uncertainty. The negative-log-likelihood is given by

$$-\ln \mathcal{L}(p, \mu, \sigma) = -\ln \left(\frac{1}{\sigma\sqrt{2\pi}} \right) + \frac{(p-\mu)^2}{2\sigma^2}, \qquad (9.35)$$

where the first term is a constant and the second term is the equation of a parabola. When $p - \mu$ corresponds to a 1σ Gaussian uncertainty the second term in Eq. (9.35) reduces to a factor of one half, and when $p = \mu$, the second term vanishes. So a change of one half in $-\ln \mathcal{L}$ from the minimum value obtained corresponds to a change in p of 1σ. In this limit of large statistics uncertainties are Gaussian by virtue of Eq. (9.34), and this result encapsulates Wilks' theorem (Wilks, 1937).

In the limit of small statistics, or where the $-\ln \mathcal{L}$ distribution is not parabolic about the minimum value it is not appropriate to attribute a change of one half to represent a $\pm 1\sigma$ boundary and Monte Carlo simulations should be used to determine the interval corresponding to the uncertainty on a parameter. In practice if the $-\ln \mathcal{L}$ distribution is almost parabolic we assume that it is valid to invoke the theorem of Wilks.

9.4.3 Adding constraints

If one is performing a fit to data given some model θ, where external information is available that may help constrain one or more parameters in the model, then it is possible to include a penalty term to the test statistic being optimised. For example, in the case of a χ^2 fit, where some parameter p has been previously measured as $p_{meas} \pm \sigma_p$, one could simply add a term to the χ^2 being minimised in the fit. This additional term would take the form

$$\chi^2_{penalty} = \frac{(p - p_{meas})^2}{\sigma_p^2}. \qquad (9.36)$$

The quantity $\chi^2_{penalty}$ would be computed for each iteration of the minimisation process and added to the usual χ^2 sum in order to incorporate external knowledge. Additional information can be included in a likelihood fit by extending the model to simultaneously fit the external information in an analogous way. It is also possible to implement constraint equations in a fit using Lagrange multipliers, for example see Edwards (1992); Silvia and Skilling (2008).

9.4.4 Fit bias and checking for self consistency

The range of fit models that one may want to apply to data varies from simple, where there are a few parameters to determine from data, to extremely complicated, where there may be tens, hundreds, or even more parameters to determine. In all cases the optimisation process used is not guaranteed to work properly, and one should take care to validate that a fit result obtained is sensible.

If one fits a sample of data, and obtains some parameter $p_{fitted} = p_0 \pm \sigma_{p_0}$, it is reasonable to ask if p_{fitted} is a good representation of the true value of the parameter p. In general, the test statistic used in the optimisation has an intrinsic bias, dependent on the sample size. In order to test for fit bias one can generate, and fit to, an ensemble of simulated data. The distributions of fitted values p_0 and σ_{p_0} obtained can be used to evaluate if the fit being performed is biased or not.

If the simulation is an accurate representation of the measurement being performed, then one will obtain a set of fitted parameters p_{fitted} for each simulated measurement in the ensemble. These should be distributed with a mean value corresponding to that used to generate the ensemble, p_{input}, and a spread representative of the uncertainty obtained from the fit. In general one can plot the so-called '*pull*' distribution

$$\mathcal{P} = \frac{p_{fitted} - p_{input}}{\sigma_p}, \tag{9.37}$$

which should be centred on zero for an unbiased measurement, with a width of one if the fit correctly extracts the uncertainty on p from data. A common reason why one might obtain a width of the distribution in Eq. (9.37) differing from one is if the simulated experiments do not allow for Poisson fluctuations that are inherent in fitting for event yields.

An additional consistency check that can be made for a maximum likelihood fit is to verify that the value of $-\ln \mathcal{L}$ obtained from fitting data is consistent with the distribution of $-\ln \mathcal{L}$ obtained from an ensemble of simulated measurements. If there is a disagreement this could indicate that there is a problem with the fit to the data, the simulation used, or both. In general the absolute value of $-\ln \mathcal{L}$ is dominated by the number of events used in the fit to data and corresponds, on average, to a sum over all events of the first term in Eq. (9.35) for fits to large samples of data.

9.5 Combination of results

The most transparent way of combining results from several different measurements of a single observable is to construct a fit model that encompasses all necessary parameters to describe all of the data used to extract information on the observable. Having done this, the data are fit using that model. This procedure can be used for situations that vary from different measurements having many parameters in common between them, to ones where only the observable of interest is common.

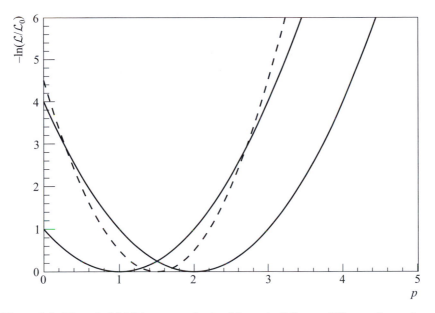

Figure 9.2 The $-\ln(\mathcal{L}/\mathcal{L}_0)$ curves obtained from (solid) two different determinations of a parameter p, with the (dashed) combined distribution shown.

In reality this approach can be complicated, and the fit validations required to understand the fit bias and performance may be impractical. If this is the case, the alternative discussed below may be useful.

Consider the situation when there are several different determinations of some parameter, each with a given likelihood or χ^2 distribution as a function of that parameter. Such a situation is not uncommon with complicated experiments as there can be more than one way to measure and observable. It is possible to combine the likelihood or χ^2 distributions of the different measurements directly. For two χ^2 distributions, the total χ^2, χ^2_{TOT}, is the sum of the individual contributions as:

$$\chi^2_{\text{TOT}} = \chi^2_1 + \chi^2_2, \tag{9.38}$$

which follows from Eq. (9.7). If the estimates of the parameter are correlated with each other then one needs to resort to using the general form of the χ^2 given by Eq. (9.22). With regard to the combination of two likelihoods $\mathcal{L}_1$ and $\mathcal{L}_2$, as each is a representation of the probability for something to happen, the combined likelihood is just the product $\mathcal{L}_1\mathcal{L}_2$. It follows that, as we usually optimise $-\ln \mathcal{L}$, the optimal value for some parameter derived from a combination of two likelihood functions is given by $-\ln \mathcal{L}_1 - \ln \mathcal{L}_2$. The minimum value of this combination corresponds to the best fit value of the parameter(s) under study, and the $\pm 1\sigma$ uncertainties can be obtained as discussed in Section 9.4.2. Figure 9.2 shows the result of combining two likelihood distributions in this way.

9.6 Template fitting

Occasionally one can encounter a problem where the output of a complicated simulation is an appropriately normalised distribution and both the normalisation and shape are a function of some underlying physical parameter of interest. In this instance, instead of having some parametric PDF given by $\mathcal{P}(x, \underline{p})$ one has a set of non-parametric PDFs, each one corresponding to a given value of the underlying parameter x. These differences in normalisation and shape can be used in a fit to constrain x in an analogous way to optimising the parameters $\underline{p}$ of $\mathcal{P}(x, \underline{p})$. For each non-parametric PDF template one can compute a value of the test statistic being minimised given a binned or un-binned sample of data. For example, if one has a binned data set, with a corresponding set of templates that have matching binning, one can compute the value of $\chi^2(x_j)$ for the jth template via

$$\chi^2(x_j) = \sum_{i=1}^{N} \frac{\left(x_i^{\text{data}} - x_{ij}^{\text{template}}\right)^2}{\sigma_{ij}^2}, \tag{9.39}$$

where the number of bins is given by N. Hence, one can compute a graph of $\chi^2(x)$ vs x which can be minimised to obtain the best fit value and uncertainty on x. An equivalent procedure can be followed with respect to performing a binned (or an un-binned) maximum-likelihood fit.

9.7 Case studies

This section discusses two situations that can be understood by applying some of the techniques discussed in this chapter. It is necessary to interpret the results of these problems in terms of the confidence regions obtained.

9.7.1 Triangulation: χ^2 approach

The problem of triangulation is common and related to a number of disciplines, for example surveying. Given a fixed baseline between two points it is possible to measure two angles, and in turn completely constrain the apex of a triangle. Such a measurement can be used to determine the position of a distant object (summit of a mountain or another notable landmark) relative to the ends of the baseline. This problem is illustrated in Figure 9.3, where the baseline is given by the distance L between point A and B, and the apex of the triangle is at point C. The internal angles of the triangle are given by α, β, and γ.

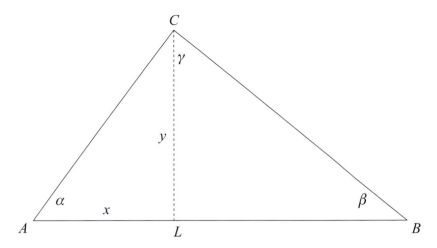

Figure 9.3 The triangulation problem: by measuring the two angles α and β, it is possible to determine the position C relative to the baseline AB.

Given that α and β are measured as the internal angles of the triangle at points A and B, respectively, it is possible to construct a χ^2 function to minimise in terms of these angles. This function is given by

$$\chi^2 = \frac{(\alpha - \hat{\alpha})^2}{\sigma_\alpha^2} + \frac{(\beta - \hat{\beta})^2}{\sigma_\beta^2}, \tag{9.40}$$

where $\hat{\alpha}$ and $\hat{\beta}$ are written in terms of the distance x along AB (as measured from A), and the perpendicular distance y from AB. Thus[4]

$$\hat{\alpha} = \arctan\left(\frac{y}{x}\right), \tag{9.41}$$

$$\hat{\beta} = \arctan\left(\frac{y}{L - x}\right). \tag{9.42}$$

In order to determine the most probable position for C in the $x - y$ plane one can scan over all reasonable values of x and y searching for the minimum value of the χ^2 given by Eq. (9.40). On determining this value it is possible to identify a contour corresponding to a change of χ^2 of one from the minimum value in order to assign an uncertainty on the position of C.

In order to further illustrate this problem, if L is a fixed distance of $1\,\text{km}$ separating two surveying points, from which one measures an angle $\alpha = (60 \pm 2)°$

---

[4] Note that when using a computer to evaluate a trigonometric function the angle input or returned is given in radians.

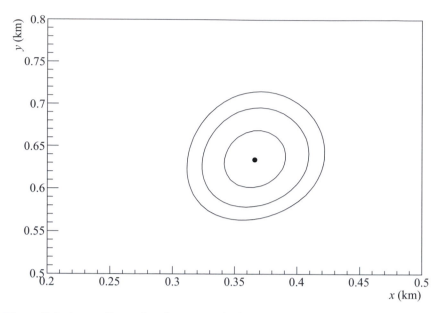

Figure 9.4 A two-dimensional contour plot indicating the most probable location of the apex of the triangle (point) in this example, along with 1σ, 3σ, and 5σ confidence regions indicated by the contour lines surrounding the point.

and $\beta = (45 \pm 2)^\circ$, then it is possible to constrain x and y using a fit based on Eq. (9.40) as described above. This yields

$$x = 0.366 \pm 0.025 \,(\text{km}), \tag{9.43}$$

$$y = 0.634 \pm 0.033 \,(\text{km}). \tag{9.44}$$

In addition to performing such a minimisation process, it is possible to scan through values of x and y in order to construct a two-dimensional contour plot indicating confidence regions related to the location of the apex of the triangle (see Figure 9.4). As the contours shown in the figure are not circular in nature, it is evident that the fitted solutions for x and y are correlated with each other. The correlation matrix obtained for this problem is

$$\rho = \begin{pmatrix} 1.00 & -0.13 \\ -0.13 & 1.00 \end{pmatrix}, \tag{9.45}$$

i.e. there is a -13% correlation between x and y.

It is possible to improve the precision on this constraint by either improving the precision on the data (α and β) or by introducing additional constraints on the problem, for example a measurement of the third angle γ, a length such as AC or BC, or indeed measurements of angles between points on the baseline AB a known distance from one of the ends of the baseline, and the point C.

A variant on this theme can be used in order to triangulate the position of a person using a mobile phone, where instead of using an angle measured between two or more nearby microwave antennae one can use the time for the phone to respond to a handshake signal sent by those antennae. For such a scenario, the triangulation problem is expressed in terms of the hypotenuse of two right-angled triangles with a known height and a known baseline which is the straight line distance between two antennae.

9.7.2 Triangulation: Bayesian approach

The problem of triangulation was discussed in the previous section in the context of a χ^2 fit-based optimisation. An alternate way to consider solving this problem is to employ Bayes' theorem as discussed in the following. If we have measurements of both α and β as illustrated in Figure 9.3, then we can use information provided by both of these measurements to compute the the probability that a given (x, y) point corresponds to the apex of the triangle given by $P([x, y]|\alpha)$ and $P([x, y]|\beta)$, starting from Bayes' theorem

$$P([x, y]|\theta) = \frac{P(\theta|[x, y])}{P(\theta)} P([x, y]), \tag{9.46}$$

where $\theta = \alpha$ or β. Prior to any measurement we can consider all values of x and y to be equally likely solutions for the apex of the triangle. If we now focus on the computation of $P([x, y]|\alpha)$, this is

$$P([x, y]|\alpha) = \frac{P(\alpha|[x, y])}{P(\alpha)} P([x, y]), \tag{9.47}$$

where we can further assume that the measurement of α is given by a Gaussian PDF with mean α and width σ_α as was the case for the previous illustration. Therefore we can write

$$P(\hat{\alpha}|\alpha) = \frac{G(\hat{\alpha}, \alpha, \sigma_\alpha)P(\hat{\alpha})}{\int G(\hat{\alpha}, \alpha, \sigma_\alpha)d\underline{x}}, \tag{9.48}$$

where we have replaced $[x, y]$ with $\hat{\alpha}$, where

$$\hat{\alpha} = \arctan\left(\frac{y}{x}\right). \tag{9.49}$$

By using Eq. (9.48) we are able to compute the posterior probability $P(\hat{\alpha}|\alpha)$ for any single point in the $x - y$ plane. We can follow a similar line of thought to obtain the posterior probability $P(\hat{\beta}|\beta)$,

$$P(\hat{\beta}|\beta) = \frac{G(\hat{\beta}, \beta, \sigma_\beta)P(\hat{\beta})}{\int G(\hat{\beta}, \beta, \sigma_\beta)d\underline{x}}, \tag{9.50}$$

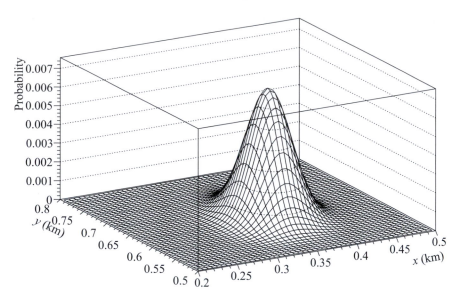

Figure 9.5 The posterior probability distribution as a function of x and y for the triangulation problem discussed in the text.

where

$$\hat{\beta} = \arctan\left(\frac{y}{L - x}\right). \tag{9.51}$$

In order to use all possible information to constrain the apex of the triangle we need to compute the product of the two posterior probabilities, that is

$$P_C = P(\hat{\alpha}|\alpha)P(\hat{\beta}|\beta). \tag{9.52}$$

Illustrating this problem using the same inputs as in Section 9.7.1, where $L = 1$ km, $\alpha = (60 \pm 2)°$ and $\beta = (45 \pm 2)°$, one obtains the following result

$$x = 0.366 \pm 0.025 \text{ (km)}, \tag{9.53}$$

$$y = 0.634^{+0.036}_{-0.034} \text{ (km)}, \tag{9.54}$$

relative to the coordinate A. The corresponding two-dimensional contour plot indicating confidence regions related to the location of the apex of the triangle is similar to the one shown in Figure 9.4. The corresponding posterior probability distribution as a function of x and y is shown in Figure 9.5. Both the χ^2 fit-based and Bayesian approaches give similar results in this case. The difference obtained for the results using these methods is in the uncertainty on y at the level of 0.003 for the upper error and 0.001 for the lower error, which for most applications is irrelevant. The fact that this interval is asymmetric is a reflection of the slightly non-Gaussian constraint obtained on the values of x and y. Whereas the χ^2 fit

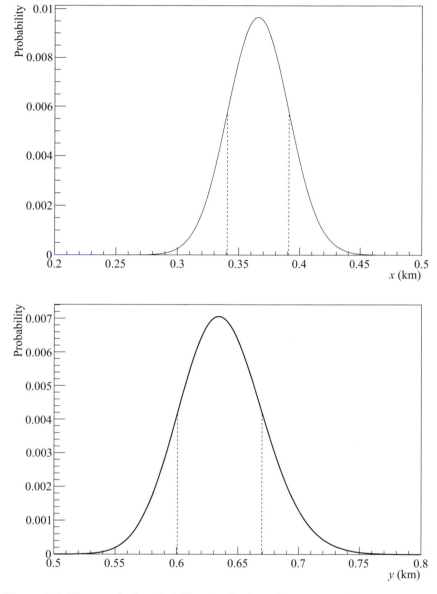

Figure 9.6 The marginal probability distributions for (top) x and (bottom) y for the triangulation problem discussed in the text. The vertical dashed lines indicate the limits of $\pm 1\sigma$ intervals interpreted as 68.3% confidence intervals distributed about the most probable values of x and y.

approach led us to interpret the constraint on x and y in terms of the change in the χ^2 from the best fit value, here we interpret these parameters in terms of the marginal probability, as can be seen from Figure 9.6. Here we have integrated $P([x, y]|\alpha, \beta)$ over y in order to determine the marginal distribution $P(x)$, and integrated it over x to obtain the marginal distribution $P(y)$. The 1σ confidence interval is constructed from these distributions by integrating contours of equal probability from the most probable result (the mode) out to the desired CL.

9.7.3 Upper limits

Chapter 7 discussed the concepts of one- and two-sided confidence intervals. The special case of one-sided confidence intervals being used to place an upper bound on the value of an observable was introduced as an upper limit. The upper limit computations discussed in that chapter are frequentist in nature. One of the limitations of a frequentist limit that was raised when discussing the unified approach (Section 7.6) and the Monte Carlo method (Section 7.7) is that of obtaining the correct coverage for a limit. It is possible to avoid this problem by computing a Bayesian upper limit using the method outlined here.

Consider a fit to data using some test statistic, either a χ^2 or some likelihood $\mathcal{L}$ (the following assumes a likelihood fit is being performed; however, similar arguments are valid for a χ^2 fit). The optimised test statistic depends on n discriminating variables $\underline{x} = \{x_1, x_2, \ldots, x_n\}$, and m parameters $\underline{p} = \{p_1, p_2, \ldots, p_m\}$, i.e. $\mathcal{L} = \mathcal{L}(\underline{x}, \underline{p})$, and one (or more) of the parameters is of interest with respect to computing an upper limit. If we consider the simplified case where we wish to obtain an upper limit on some parameter p_j determined from a fit to data, for each value of p_j we must fit the data (p_j held constant, and all $p_i, i \neq j$ allowed to vary in the fit) in order to obtain the optimised test statistic for a given value of p_j. The set of parameters determined for this scenario can be used to plot p_j against $\mathcal{L}$ as shown in Figure 9.7.

This particular parameter is assumed to be meaningful for $p_j \geq 0$, and the figure is truncated for un-physical values. The modal value $p_{j,\mu}$ is less than 3σ from zero and, as a result, we may like to compute an upper bound on this parameter at a some confidence level (for example 90% CL). We can use Bayes' theorem to compute the upper limit by noting

$$0.9 = \frac{\displaystyle\int_{p_j=-\infty}^{p_{j,UL}} \mathcal{L}(\underline{x}, \underline{p})\theta(p_j)dp_j}{\displaystyle\int_{p_j=-\infty}^{\infty} \mathcal{L}(\underline{x}, \underline{p})dp_j}, \tag{9.55}$$

where $\theta(p_j)$ is the prior. There are two balancing concerns for the prior, the first is that p_j is physical for $p_j \geq 0$, thus we would expect $\theta(p_j) = 0$ for $p_j < 0$. The second concern is the shape of the prior for physically allowed values of p_j. If we are completely ignorant about what value p_j may take, then we can assume that $\theta(p_j) =$ constant for $p_j \geq 0$ and absorb the constant into the likelihood. Thus

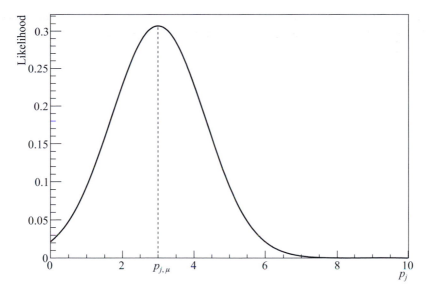

Figure 9.7 The likelihood as a function of p_j. The most probable value of p_j is indicated by $p_{j,\mu}$ (dashed line).

Eq. (9.55) reduces to

$$0.9 = \frac{\int\limits_{p_j=0}^{p_{j,UL}} \mathcal{L}(\underline{x}, \underline{p}) dp_j}{\int\limits_{p_j=0}^{\infty} \mathcal{L}(\underline{x}, \underline{p}) dp_j}, \qquad (9.56)$$

which can be solved numerically if an analytic solution is not possible. As was mentioned previously, coverage can be an issue unless it is explicitly verified when computing an upper limit using frequentist methods. As can be seen from Eq. (9.56), by construction the coverage obtained for the upper limit on $p_{j,UL}$ is the desired one by definition.

If we do have some a priori information about the value of p_j, for example that this parameter has been constrained by some previous measurement, we are able to incorporate that information into $\theta(p_j)$ so that the prior would not be allowed to become incompatible with the existing measurement. As with previous discussions on the matter, the actual result obtained should not depend strongly on the value of the prior used.

Example. Consider the situation where the likelihood for some measurement as a function of a parameter p is given by a Gaussian distribution with a mean value $\mu = 3.0$, and a width $\sigma = 1.3$. This corresponds to the likelihood distribution shown in Figure 9.7. The Gaussian probability tables in Appendix E can be used to determine that the the normalisation integral has the value of 0.9896 (corresponding to the one sided integral of $\mathcal{L}$ from $z = -2.31$ to $+\infty$). Therefore, for a 90% CL upper limit we require

$$0.8906 = \int_{p_j=0}^{p_{j,UL}} G(p_j, 3.0, 1.3)dp_j. \tag{9.57}$$

$$= \int_{p_j=0}^{3.0} G(p_j, 3.0, 1.3)dp_j + \int_{p_j=3.0}^{p_{j,UL}} G(p_j, 3.0, 1.3)dp_j, \tag{9.58}$$

$$= 0.4896 + \int_{p_j=3.0}^{p_{j,UL}} G(p_j, 3.0, 1.3)dp_j. \tag{9.59}$$

Thus $p_{j,UL} = 4.68$ at 90% CL as the second integral has to equal 0.401. There is a 0.013 error in the value of $p_{j,UL}$ from the binning used for Table E.8.

9.8 Summary

The main points introduced in this chapter are listed below.

(1) Having defined the optimisation technique that you want to use, it is possible to develop a model of data defined by some test statistic, and to determine the optimal match of model parameter to the data. The gradient descent method is one possible optimisation technique.
(2) Test statistics that were discussed in this section include the use of χ^2 (Section 9.2) and maximum likelihood (Section 9.4) methods. The linear least-squares approach was also discussed (Section 9.3), as a variant of the χ^2 technique.
(3) It is possible to scan through parameter space in order determine the best fit to data. See Section 9.2.1 for an example of this.
(4) In addition to discussing several examples of how these techniques can be used with data, there is some discussion on how to combine results (Section 9.5).

The Introduction includes a description of how one can test Ohm's law and measure the half-life of a radioactive isotope. The techniques required to understand in detail how such an experiment can be performed have now all been discussed in earlier parts of this book. You may wish to revisit Sections 1.2 and 1.3 in order to review those examples and reflect on your understanding of these techniques. The interested reader may also wish to refer to the books by Barlow (1989); Cowan (1998); Davidson (2003); Edwards (1992); James (2007); Kendall and Stuart (1979) for a more advanced discussion of this topic.

Exercises

9.1 When might it be useful to perform a fit to data?

9.2 Explain the term 'number of degrees of freedom' and how this relates to the number of data N in a data set.

9.3 Given the measurements $x_1 = 1.2 \pm 0.3$, and $x_2 = 1.8 \pm 0.3$, approximate the mean value and uncertainty on the average value of x obtained by performing a χ^2 scan from 1.0 to 2.0 in steps of 0.1. Compare the results obtained with those from a weighted average (see Chapter 6, Exercise 6.8).

9.4 Assuming that $y = ax^2 + b$, use the method of least squares to derive the values of a and b. Assume that the uncertainties on each of the data points is the same.

9.5 Use the method of least squares, assuming the relationship $y = ax + b$, to determine the values and errors associated with a and b given the data below, assuming that the uncertainty $\sigma_y = 0.1$.

x	1.1	1.5	2.0	3.1	4.2	5.0
y	2.0	2.9	4.2	6.0	8.0	10.0

9.6 By performing a χ^2 scan, estimate the average value of the following measurements of the quantity S: 0.655 ± 0.024, 0.59 ± 0.08, 0.789 ± 0.071. Hint: you might like to assume a bin spacing of 0.01 and consider the range $S = 0.6$ to $S = 0.7$.

9.7 Given two measurements of some observable x,

$$x_1 = 1.3 \pm 0.1, \tag{9.60}$$

$$x_2 = 1.1 \pm 0.2, \tag{9.61}$$

estimate the best fit for the average value of x_1 and x_2 by performing a χ^2 scan in steps of 0.05 from $x = 1.1$ to $x = 1.4$, and estimate the the error on the average value. Note the χ^2, and the corresponding $P(\chi^2, v)$ for your result, and comment if this is reasonable.

9.8   If a $-\ln \Delta \mathcal{L}$ curve is given by $1.1(x - 1.3)^2$, what is the central value and uncertainty on the best fit result obtained?

9.9   If a $-\ln \Delta \mathcal{L}$ curve is given by $0.5(x + 2)^2$, what is the central value and uncertainty on the best fit result obtained?

9.10 If a $-\ln \Delta \mathcal{L}$ curve is given by $(x - 1)^2$, what is the central value and uncertainty on the best fit result obtained?

9.11 Hooke's law is given by $F = -kx$, where F is force in Newtons, and x is the elastic extension in centimetres. Given the following data for a spring under load, determine the best fit value for the constant of proportionality k, assuming the uncertainty on x_i is constant.

F	1.0	2.0	3.0	4.0	5.0
x	0.4	1.1	1.5	2.1	2.5

9.12 Ohm's law is given by $V = IR$. Given that current is measured by a precise device, and that $\sigma_V(I)$ is a constant, determine the general expression for calculating R for a data sample using a least squares approach. This can be applied to the example discussed in Chapter 1.

10

Multivariate analysis

Consider a data sample Ω described by the set of variables $\underline{x}$ that is composed of two (or more) populations. Often we are faced with the task of trying to identify or separate one sub-sample from the other (as these are different classes or types of events). In practice it is often not possible to completely separate samples of one class A from another class B as was seen in the case of likelihood fits to data. There are a number of techniques that can be used in order to try and optimally identify or separate a sub-sample of data from the whole, and some of these are described below. Each of the techniques described has its own benefits and disadvantages, and the final choice of the 'optimal' solution of how to separate A and B can require subjective input from the analyst. In general this type of situation requires the use of multivariate analysis (MVA).

The simplest approach is that of cutting on the data to improve the purity of a class of events, as described in Section 10.1. More advanced classifiers such as Bayesian classifiers, Fisher discriminants, neural networks, and decision trees are subsequently discussed. The Fisher discriminant described in Section 10.3 has the advantage that the coefficients required to optimally separate two populations of events are determined analytically up to an arbitrary scale factor. The neural network (Section 10.4) and decision tree (Section 10.5) algorithms described here require a numerical optimisation to be performed. In this context the optimisation process is called training, and having trained an algorithm with a set of data one has to validate the solution. Having discussed several classifier algorithms, the concepts of bagging and boosting are described as variants on the training process. In the following it is assumed that the data Ω contain only two populations of events. These populations are either referred to as A and B, or as signal and background depending on the context. It is possible to generalise these approaches to an arbitrary number of populations.

10.1 Cutting on variables

An intuitive way to try and separate two classes of events is to *cut* on the data to select interesting events in a cleaner environment than exists in the whole data set. For example, if you consider the case where the data set Ω contains two types of events A and B that are partially overlapping, one possible solution to the problem of separating A from B is to select the events that satisfy $A \setminus B$. If $C = A \setminus B \neq \emptyset$, then this will be a pure sample of interesting events.[1] The pertinent questions are: (i) what sacrifice has been made in order to obtain C, and (ii) would it have been possible to reject less data and obtain a more optimal separation of A and B so that we can further study a subset of the data?

What do we mean by making a cut on the data? Consider the data sample Ω above which contains two classes of events: A and B, each of which is described by discriminating variables in n dimensions. If we cut on the value of one or more of the dimensions, then we decide to retain an event e_i that passes some threshold

$$P(e_i \in A) > 0 \qquad\qquad (10.1)$$

and would decide to discard an event if

$$P(e_i \notin A) \text{ is significant.} \qquad\qquad (10.2)$$

There is an element of subjectivity in the second condition. We can think of a cut in some variable x as a binary step function $f(x)$, where in the case of a positive step we may write

$$f(x) = 1 \text{ for } x > X_0, \qquad\qquad (10.3)$$

$$f(x) = 0 \text{ elsewhere,} \qquad\qquad (10.4)$$

and for a negative step, we change the inequality from $>$ to $<$. In order to optimise the cut on x we need to determine what it is we aim to achieve. If we assume our signal events are those of class A, then it follows that we would like to retain as many events of type A as possible, while discarding as many events of type B as possible. If $A \cap B = \emptyset$, then it is possible to determine X_0 by inspection of the distributions. If, however, $A \cap B \neq \emptyset$, we need to choose what we mean by the term *optimally separating A and B*.

The following lists some of the possible ways to determine X_0 for *optimal* separation for a given test statistic.

(1) If it is of paramount importance to obtain a pure sample of A with no contamination or dilution from B, then we define X_0 in such a way that satisfies

---

[1] Purity is defined as the fraction of signal (or interesting events) in a sample of data.

$C = A \setminus B \neq \emptyset$ with as many events passing into C as possible. Practically this usually is achieved at a significant cost in statistics and so will probably not be a sensible criteria for optimisation for many of the situations encountered.

(2) We can introduce the notion of the ***significance*** S of the signal content (amount of A) in Ω relative to the background (amount of B). Then we can choose the value of X_0 that results in the greatest significance of signal. A common definition of significance is the test statistic

$$S = \frac{N_S}{\sqrt{N_S + N_B}},$$

(10.5)

where N_S is the number of signal events, and N_B is the number of background events that pass a given cut with cut-value $x = X_0$. The motivation for this definition of significance is that we want to compare any hint of a signal found to the statistical uncertainty on the number of events in the data. The underlying logic is that we want to be able to minimise any incorrect claim of a signal that would arise from statistical fluctuations in the data sample. As a result if we compute a numerical value for S, we may say that the expected significance for a given cut is $S\sigma$, assuming that the denominator corresponds to a Gaussian uncertainty on the total number of observed events.

(3) If we are searching for an effect that is expected to be absent from the data, then we may want to optimise in such a way that we minimise the uncertainty on the background estimation (or number of events of type B that will remain in the sample), as this will dominate the uncertainty we obtain on the possible presence of a signal, and hence on any limit we are able to place that rules out the effect we seek. A possible test statistic to use in this case is the signal-to-background ratio N_S/N_B.

The previous discussion with regard to making cuts has been based on a single dimension. In the case that all relevant dimensions $\underline{x}$ in Ω are un-correlated, it is sufficient (and efficient) to optimise the cut values $\underline{X_0}$ one dimension at a time. The values of $\underline{X_0}$ obtained through such a procedure would be optimal. The more general situation encountered is when two or more dimensions are correlated. For such cases one would ideally like to simultaneously optimise the values of $\underline{X_0}$; however, this is often not practical in terms of time or resources.[2] A possible alternative to this is to iteratively optimise the values of $\underline{X_0}$ one dimension at a time. If on subsequent iterations of the optimisation the value of X_0 obtained for a given dimension does not change appreciably, then you will have obtained the cut

---

[2] The number of iterations required to simultaneously optimise m dimensions scales as the number of iterations for one dimension raised to the power of m. This is referred to as the curse of dimensionality as originally noted by Bellman (1961).

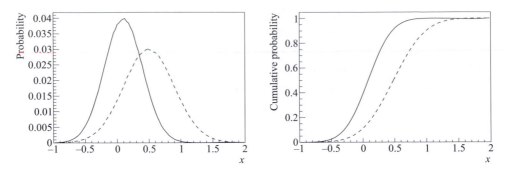

Figure 10.1 The distribution (left) in x of simulated (solid) signal and (dashed) background events, and (right) the cumulative probability distributions summing up from left to right for the cut based optimisation example described in the text.

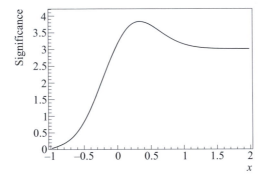

Figure 10.2 The significance computed as $N_S/(\sqrt{N_S + N_B})$ for the cut based optimisation example described in the text.

value for this dimension. In practice it may take several iterations to achieve this when two or more dimensions are correlated.

Example. Given a sample of data with an expected number of 100 signal events over a background of 1000 events, what is the optimal cut value to maximise the significance $S = N_S/(\sqrt{N_S + N_B})$? In order to determine this, we use 10^6 simulated data events for signal and background with known mean and widths that correspond to that expected in the data. These distributions are shown in Figure 10.1, where the signal and background are distributed according to Gaussian PDFs with means of 0.1 and 0.5 and widths of 0.3 and 0.4, respectively. It can be seen that in this case there is a trade off from allowing background to pass the cuts, while retaining a reasonable signal efficiency. The figure also shows the cumulative probability distributions for signal and background, which is equivalent to the efficiency of selecting events for a cut X_0 that rejects higher values of x. The resulting significance distribution for this situation is shown in Figure 10.2

where one can see that $X_0 = 0.34$ would provide optimal separation between signal and background using this method. The significance has a maximum value of 3.8σ for this value of X_0. On further inspection of the figure one can see that for larger values of X_0, there is a drop in significance arising from an increase in background. For smaller values of X_0 there is a drop in significance as signal is removed that would otherwise contribute to a measurement.

10.1.1 *Optimisation of cuts as a precursor to further analysis*

It should be noted that if the aim of a cut-based selection of events is to subsequently use those events with a more complicated algorithm such as a fit-based optimisation as discussed in Chapter 9, or with an MVA like those described in the remainder of this chapter, then it doesn't make sense to optimise cut values as described above. The end result of your experiment will be the result of analysis with that more sophisticated technique – and so it is that which should be used to determine what is the *optimal* measurement to make. If one does optimise with a cut-based approach, and then performs a more sophisticated data analysis, the end result will generally be less precise than if one applied a loose set of cuts, then performed the subsequent data analysis. It is important to stress that control samples of data or simulated data should be used in the optimisation process so that the result you obtain is not biased.

10.2 Bayesian classifier

The scenario of hypothesis comparison described in Section 8.4 can be extended generally to a classification problem, given some data Ω that can be tested against a set of classifications given by H, where the ith classification is given by H_i. For each event ω_j in the data set we can compute the probability $P(\omega_j|H_i)$ that the event is of the ith classification. The most probable hypothesis is given by

$$P_{max}(\omega_j|H_i) = max[P(\omega_j|H_i)], \tag{10.6}$$

i.e. the largest value of $P(\omega_j|H_i)$ for all i is used to identify the classification for an event. Such a classifier is referred to as a Bayesian classifier.

Example. Consider the situation where one is interested in identifying three categories of event: (i) interesting I and in need of detailed study, (ii) possibly interesting PI at some level, and (iii) not interesting NI. One can compute

$$P_I = P(\omega_i|I), \tag{10.7}$$

$$P_{PI} = P(\omega_i|PI), \tag{10.8}$$

$$P_{NI} = P(\omega_i|NI). \tag{10.9}$$

If the largest probability for event ω_i is P_I, one will classify the event as interesting and in need of further study. Similarly if the largest probability is P_{PI} or P_{NI}, the event would be classified as possibly interesting or not interesting, respectively. A more specific example utilising Bayesian classifiers is discussed in Section 10.7.2.

10.3 Fisher discriminant

Fisher's linear discriminant (or *Fisher discriminant*) is a linear combination of the variables $\underline{x}$ to form a single classifier output $\mathcal{O}$ given by (Fisher, 1936)

$$\mathcal{O} = \sum_{i=1}^{n} \alpha_i x_i + \beta, \tag{10.10}$$

$$= \underline{\alpha} \cdot \underline{x} + \beta. \tag{10.11}$$

The sum is over the number of dimensions n in the classification problem. In order to make use of Eq. (10.11) in practice we need to determine the weight coefficients α_i, or equivalently the weight vector $\underline{\alpha}$. The value of β does not affect the separation between data types, it adjusts the overall central value of the resulting Fisher distribution, and in the following discussion this parameter will be set to zero.

Given the data set Ω and the knowledge of which elements in Ω are of class A and which are of class B we can compute the mean and variance of $\underline{x}$ for the two classes. These are $\underline{\mu}_{A,B}$ and $\sigma^2_{A,B}$ where the subscript indicates the event type. Using Eq. (10.11) we can also compute the mean M and variance Σ^2 of the Fisher distributions for the two classes of data

$$M_{A,B} = \alpha^T \mu_{A,B} = \sum_i \alpha_i \mu_{A,B}, \tag{10.12}$$

$$\Sigma^2_{A,B} = \alpha^T \sigma^2_{A,B} \alpha = \sum_i \sum_j \alpha_i \sigma_{ij\,A,B} \alpha_j, \tag{10.13}$$

where we now revert to matrix notation to avoid having to explicitly write out the summations involved. In order to maximise the separation between A and B we want to maximise the difference between M_A and M_B, while at the same time minimise the sum of the variances of the two output distributions. These requirements are expressed in the ratio

$$J(\alpha) = \frac{[M_A - M_B]^2}{\Sigma^2_A + \Sigma^2_B}, \tag{10.14}$$

where the squared sum of the mean values of the Fisher distribution for the two classes is

$$[M_A - M_B]^2 = \left[\sum_{i=1}^{n} \alpha_i (\mu_A - \mu_B)_i\right]\left[\sum_{j=1}^{n} \alpha_j (\mu_A - \mu_B)_j\right], \quad (10.15)$$

$$= \sum_{i,j=1}^{n} \alpha_i (\mu_A - \mu_B)_i (\mu_A - \mu_B)_j \alpha_j, \quad (10.16)$$

$$= \alpha^T B \alpha, \quad (10.17)$$

where the matrix B is introduced to represent the separation *between* the classes of events based on mean values. The the sum of the Fisher distribution variances is

$$\Sigma_A^2 + \Sigma_B^2 = \alpha^T \sigma_A^2 \alpha + \alpha^T \sigma_B^2 \alpha, \quad (10.18)$$

$$= \alpha^T W \alpha, \quad (10.19)$$

where the matrix W is the sum of covariance matrices *within* the classes. Thus, we find

$$J(\alpha) = \frac{\alpha^T B \alpha}{\alpha^T W \alpha}. \quad (10.20)$$

The optimal separation between classes A and B can be found by minimising J with respect to the weight coefficients α, therefore by satisfying the condition

$$\frac{\partial J(\alpha)}{\partial \alpha} = 0. \quad (10.21)$$

One can show (for example, see Cowan (1998)) that the maximum separation is found when

$$\alpha \propto W^{-1}(\underline{\mu}_A - \underline{\mu}_B), \quad (10.22)$$

so we are able to compute the weights α if we are able to determine the mean values $\mu_{A,B}$, and invert the matrix W. As the coefficients are determined up to some proportionality, we don't have a unique solution for the set of weights, but have a family of related solutions. This method implicitly assumes that the matrix W can be inverted. If W is singular, then one either has to change the input dimensions to produce a non-singular W matrix, or alternatively use a different classification method.

If we so wish, we can extend the form of Eq. (10.11) by scaling or offsetting the input data to lie within a specified range. Furthermore it is possible to scale or offset the computed $\mathcal{O}$ as desired if you want to relocate the mean value or change the range over which the classifier outputs are computed for the data. On doing this

Multivariate analysis

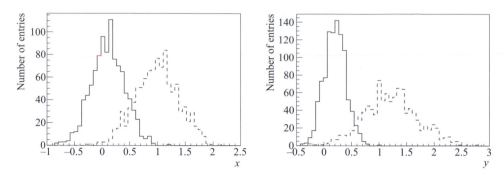

Figure 10.3 Distributions of (left) x and (right) y for (solid) signal and (dashed) background events for the example described in the text.

the separation between types A and B will remain optimal as defined by the Fisher algorithm.

Example. Consider the situation where we have a data sample comprising two types of events: signal (S) and background (B), each described in two dimensions that are independent: x and y. We want to compute a set of Fisher discriminant coefficients α to separate out S from B so that we can further analyse a clean sample of the signal events. From the data sample, we are able to compute

$$\mu_S = \begin{pmatrix} 0.1 \\ 0.2 \end{pmatrix}, \text{ and } \sigma_S = \begin{pmatrix} 0.3 & 0.0 \\ 0.0 & 0.2 \end{pmatrix}, \tag{10.23}$$

for the signal, and

$$\mu_B = \begin{pmatrix} 1.0 \\ 1.2 \end{pmatrix}, \text{ and } \sigma_B = \begin{pmatrix} 0.4 & 0.0 \\ 0.0 & 0.5 \end{pmatrix}, \tag{10.24}$$

for the background. Here σ_A and σ_B contain the standard deviations of the independent variables x and y, and are not covariance matrices. The distributions of the signal and background data are shown in Figure 10.3. There are regions of the signal that are background free, and similarly there are regions of the background data that are signal free. The objective is to obtain an optimal separation between the two classes of events.

Given this information we can compute the difference in mean values of S and B as

$$(\mu_S - \mu_B) = \begin{pmatrix} -0.9 \\ -1.0 \end{pmatrix}, \tag{10.25}$$

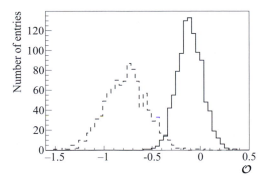

Figure 10.4 The Fisher discriminant output distribution $\mathcal{O}$ for (solid) signal and (dashed) background events for the example described in the text.

and W is given by[3]

$$W = \begin{pmatrix} 0.25 & 0 \\ 0 & 0.29 \end{pmatrix}, \qquad (10.26)$$

where the off-diagonal terms are zero as x and y are uncorrelated for both signal and background. From Eq. (10.22) we can determine the weight vector up to some arbitrary scale factor to be

$$\alpha = \begin{pmatrix} -3.6 \\ -3.45 \end{pmatrix}. \qquad (10.27)$$

Figure 10.4 shows the output Fisher distribution $\mathcal{O}$ obtained using the weights computed for this example. The separation between signal and background distributions in terms of $\mathcal{O}$ is better than the separation either with x or with y when one compares with the distributions in Figure 10.3. The signal distribution appears on the right-hand side of the figure as a result of the convention adopted in Eq. (10.22) where background means are subtracted from signal ones.

10.3.1 Choice of input variables

Often we have a choice of input variables or dimensions that we want to use to separate between classes of events. Some common sense should be used when doing this, as for example if you introduce a dimension where A and

---

[3] The original σ matrices provided in this example contain the standard deviations of the data, and the discriminating variables x and y are un-correlated. Hence the individual standard deviations need to be squared to obtain the covariance matrix, and using this one can then construct W. This step is not required if one starts from the two covariance matrices.

B are almost completely overlapping with similar distributions, that dimension will have essentially no weight in the final Fisher discriminant that you compute. In turn you may decide that it is not worth including that variable in your classifier.

A corollary of the method is that if the mean value of the distribution of events in a given dimension is the same for both classes *A* and *B*, but the shapes of the two distributions are rather different, by definition the corresponding weight α_i will be zero (this follows from Eq. (10.22)). In such cases it makes sense to transform the variable somehow in order to make sure that the mean values of the distributions for *A* and *B* are different. One possible way to do this if you have a common mean value for types *A* and *B* and events are distributed differently for the two types, is to fold the data about the mean value and use the resulting distributions as an input to the Fisher discriminant. The resulting distributions will not be symmetric about a common mean, and the variable will in turn have a contribution to the separation between the two classes of event.

10.4 Artificial neural networks

There are many variants on the concept of artificial neural networks. These are all built upon complex structures assembled from individual perceptrons, see Section 10.4.1. The type of neural network that is most commonly used in physics applications is that of a multi-layer perceptron (MLP), see Section 10.4.2. The MLP is an ensemble of layers of perceptrons used in order to try and optimally separate classes of events. Typically there are *n* dimensions input to the network and only a single output; however, it is also possible to configure a network with multiple outputs. Only single-output MLPs are discussed here. An important, and often overlooked, aspect to the use of neural networks is that of validation. After describing the MLP, there is a discussion on training methods in Section 10.4.3, and the issue of validation is discussed in Section 10.4.4. More information on this topic can be found in a number of books (see, for example, Hastie, Tibshirani and Friedman, 2009; MacKay, 2011; Rojas, 1996).

10.4.1 Perceptrons

The fundamental building block of a neural network is the perceptron. The ***perceptron*** is the algorithmic analogy of a neuron. There are *n* inputs to the perceptron, these provide an impulse for it to react to. The perceptron has a predefined action which in turn performs some function and finally gives a response in the form of an output (see Figure 10.5). The simplest type of perceptron is a binary

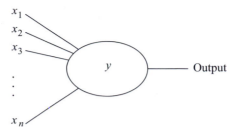

Figure 10.5 A single perceptron with n input values, an activation function y, and a single output.

threshold perceptron. This takes an n-dimensional input in the form of an event e_i described by the vector $\underline{x}_i$. Given $\underline{x}_i$, the perceptron is used to compute some response $\mathcal{O}$ using a so-called **activation function** y. The binary threshold perceptron algorithm is

$$y_i = \underline{w} \cdot \underline{x}_i + b,$$
$$\mathcal{O} = 1 \text{ if } y_i > 0,$$
$$= 0 \text{ otherwise.} \tag{10.28}$$

If y_i is above threshold for a given event (tuned by the parameter b) the response is one, and if it is below threshold the response is zero. The vector $\underline{w}$ corresponds to the set of weights used to separate classes of events from each other, and b is a constant offset used to tune the binary perceptron's threshold value. If we think about what the perceptron is actually doing, one can see that we are defining a plane in an n-dimensional space as $\underline{w} \cdot \underline{x}_i + b$, and then accepting all events that occupy space on one side of this plane. The events on the other side of the plane are rejected. In order to optimally select interesting events using a single perceptron we need to determine the parameters $\underline{w}$ and b. So for each perceptron there are $n + 1$ weights (or n weights if you set b to zero) to determine. Training is discussed in more detail in Sections 10.4.3 and 10.4.4.

The n-dimensional binary threshold perceptron is equivalent to the n-dimensional cut-based event selection described in Section 10.1 applied to a Fisher discriminant. Both algorithms are making a cut in the problem space, and one recognises the similarity between the activation function $\underline{w} \cdot \underline{x}_i + b$, and Eq. (10.11).

The function given in Eq. (10.28) is called the activation function of the binary threshold perceptron. In practice we are not restricted to a single type of activation function. The following types of activation function are commonly used.

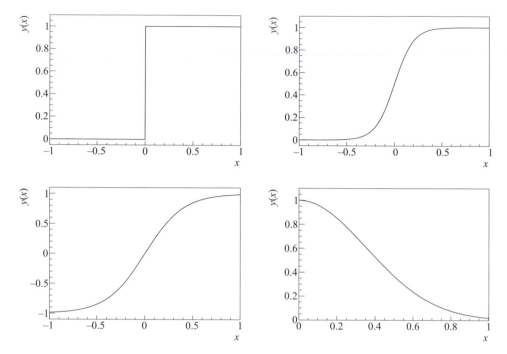

Figure 10.6 Example distributions of the (top left) binary, (top right) sigmoid, (bottom left) hyperbolic tangent, and (bottom right) radial activation functions.

- The n-dimensional binary threshold function given by Eq. (10.28).
- A sigmoid (or logistic) function given by

$$y = \frac{1}{1 + e^{\underline{w} \cdot \underline{x}_i + \beta}}, \qquad (10.29)$$

  which varies smoothly with output values in the range of 0 to $+1$.
- The hyperbolic tangent: $y = \tanh(\underline{w} \cdot \underline{x}_i)$ which varies smoothly with output values in the range -1 to $+1$.
- The radial function: $y = e^{-\underline{w} \cdot x_i^2}$ which varies smoothly between 0 and 1.

Figure 10.6 shows example distributions of the aforementioned activation functions. By using a smoothly varying activation function, as opposed to the binary threshold function described previously, we are able to finely tune the decision as to whether an event is signal like or not in terms of a continuous variable. Another way of thinking about this is that it is possible to consider an event to be a little like signal or background, without having to make a hard judgment as to whether the event is definitely signal or background. This can be useful when sample distributions overlap in data as is often the case. One can think of the use of a continuous activation function as a blurred cut in parameter space for events that lie on the

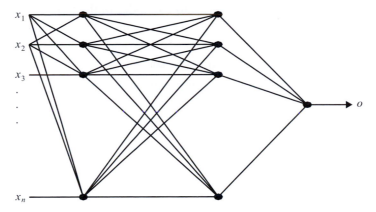

Figure 10.7 A multi-layer perceptron with n input values, one hidden layer of n nodes, and a single output.

boundary between classification as type A or type B, compared with the hard cut that would be imposed by the binary threshold function. This is equivalent to the probabilistic treatment of events in a likelihood fit.

10.4.2 Multi-layer perceptron

A **neural network** is a combination of perceptrons, each with n inputs. It is possible to have a single perceptron to govern the output of the network, which would combine the decisions made by each of the input nodes into a single output. Usually the output would be a continuous number between either zero and one or -1 and $+1$ to indicate if an event e_i was signal like ($\mathcal{O} = +1$) or not ($\mathcal{O} = -1$ or 0 depending on the activation function).

In general, a **multi-layer perceptron** is more complicated than this picture, and there will be a single input layer connected to the output node via one or more hidden layers. Figure 10.7 shows an MLP with n input nodes, one hidden layer of n nodes, connecting to a single output node. Each of the input nodes has n inputs, and the output of each of these nodes is transmitted to all of the nodes in the next layer. As each perceptron has at least n weight parameters to determine, if there are several hidden layers and n is large, the number of parameters to determine rapidly increases. For m perceptrons, in an input layer, each with an n-dimensional input, feeding o perceptrons in a hidden layer, and a single output perceptron, then the number of parameters to determine in order to compute the output of the MLP is: $n \times m$ for the input layer, $m \times o$ for the hidden layer, and o for the output node. So there would be a total of $(n + o) \times m + o$ parameters to determine. This assumes that the activation function for each node depends only on factors of

$\underline{w} \cdot \underline{x}_i$. So if one has ten inputs ($n = 10$), to ten nodes in the input layer ($m = 10$), with a hidden layer of ten nodes ($o = 10$), and a single output layer, the number of weights to compute is 210. Such a network would be described as having a $10 : 10 : 1$ configuration in shorthand. Even for a $5 : 5 : 1$ MLP with five inputs, one would have 55 parameters to determine. Such flexibility in the configuration of a network means that a lot of care needs to be taken to ensure that the trained set of *optimal* weights is not fine tuned on fluctuations in training samples. A method of determining the weight parameters is discussed in more detail in Section 10.4.3, and Section 10.4.4 discusses the importance and main issues of validating the computed weights.

10.4.3 Training an MLP

The process of determining the weight parameters for neural network is called *training*. There are several steps involved in training a MLP which are as follows.

(1) Define an algorithm to assign an error to a given set of weights.
(2) Define the procedure for terminating training, based on the computed error, or other information.
(3) Guess an initial set of weights to test the classification process.
(4) Evaluate the error defined in step (1) for a given set of data containing (preferably) equal numbers of target types for signal and background.
(5) Determine a new set of weights based on mis-classified events.
(6) Iterate the previous two steps until the convergence criteria defined in step (2) has been reached.
(7) Validate the weights obtained via this procedure (see Section 10.4.4).

The case of a single perceptron

The *error* assignment for a single perceptron is based on the ability to correctly classify if an event e_i is of the appropriate type. For example, signal events should be classified as signal, and background events should be classified as background.

If the signal classification (class A) is type $= 1$, and the background classification (class B) is assigned type $= 0$ for an event e_i, then we can define an error on the output of a perceptron ϵ_i as

$$\epsilon_i = \frac{1}{2}(t_i - y_i)^2, \tag{10.30}$$

where t_i is the true target type for the event, and y_i is the output of the perceptron. The value of y_i computed for an event will depend on the set of weight vectors used in the computation, and on the form of the activation function chosen for the perceptron. The mis-classification of the event is given by $t_i - y_i$; however, we want to be able to sum up error terms, and so it is conventional to square this difference to maintain a positive definite quantity. Similarly the factor of $1/2$ is also conventional.

If there are N events in the data sample Ω, then the total error from a single perceptron will be given by

$$E = \sum_{i=1}^{N} \epsilon_i \qquad (10.31)$$

$$= \frac{1}{2} \sum_{i=1}^{N} (t_i - y_i)^2. \qquad (10.32)$$

Having computed the error on the event classification it is desirable to be able to compute a new set of weight vectors that are closer to the optimal set than the initial guess. If the initial weight vector is $\underline{w}_m$, then we want to compute a new weight vector

$$\underline{w}_{m+1} = \underline{w}_m + \Delta \underline{w}, \qquad (10.33)$$

such that $\underline{w}_{m+1}$ is closer to the minimum of the total error function than $\underline{w}_m$. If we consider the E versus $\underline{w}$ parameter space, then we can estimate the derivative of E with respect to $\underline{w}$ as

$$\frac{\Delta E}{\Delta \underline{w}} \simeq \frac{\partial E}{\partial \underline{w}}, \qquad (10.34)$$

so

$$\Delta E \simeq \Delta \underline{w} \frac{\partial E}{\partial \underline{w}}. \qquad (10.35)$$

If we choose $\Delta \underline{w}$ such that it depends on the rate of change of E with respect to the weight vector we can ensure that we take a small step toward the minimum if

$$\Delta \underline{w} = -\alpha \frac{\partial E}{\partial \underline{w}}, \qquad (10.36)$$

where α is a small positive parameter called the ***learning rate***. Thus the total change in error ΔE given by

$$\Delta E \simeq -\alpha \left(\frac{\partial E}{\partial \underline{w}} \right)^2, \qquad (10.37)$$

which is always a negative quantity by construction. The functional form of E is given by Eq. (10.32), so once the activation function is defined, hence the $\underline{w}$ dependence of E has been chosen, it is possible to compute $\Delta \underline{w}$ and hence ΔE for a data sample. This method of iteratively computing weight vectors is often called the gradient descent method or the Δ rule, and was previously encountered when discussing optimisation of fit model parameters in Section 9.1.1.

Back propagation: training an MLP

When one moves from a single perceptron to an MLP, the error assignment algorithm is more complicated. One has to assign some importance to different contributions to the final network output, and where necessary to work back from the output layer to the input layer to modify the choice of weights. Back propagation is a generalisation of the gradient descent algorithm discussed above. The weight determination for the input layer of perceptrons is based on Eq. (10.33), where once again the objective is to minimise the total error rate E. Each perceptron in the network contributes an error rate corresponding to Eq. (10.32).

As with the case of training a single perceptron, having determined the error for an ensemble of events, given an initial assumed set of weights, one can iterate and estimate a new set of weights. This process follows an analogous procedure to that outlined above. This method is a generalisation of the Δ rule, so again it works on the concept of error minimisation through gradient descent. Detailed descriptions of the back propagation method can be found in a number of texts; for example, see MacKay (2011) and Rojas (1996).

10.4.4 Training validation for a neural network

Training validation is discussed as a subsection in its own right to highlight the importance of this topic. It is not sufficient to assume that a computed set of weights for a network is correct. Having obtained what is assumed to be a reasonable set of weights, it is necessary to perform cross-checks to ensure that the solution is not tailored to statistical fluctuations in the data used to compute them. In general there are many local minima that could be found through the minimisation of E with respect to the weight parameters – so how can one determine if the minimum obtained is really the global minimum, or if it is one of the local minima?

The problem arises as the MLP with a given set of weights w has a total classification error E as computed for some training data sample Ω_{train}. This training sample is a reference where target types of each event, either as class A or as class B are known beyond doubt. In practice we will want to apply the MLP to a classification problem using a different data sample comprising real data Ω_{data} where the target type is not certain. How do we know that the MLP will behave reasonably when applied to Ω_{data}? If we have sufficient training data then we can construct a statistically distinct set Ω_{validate} that is equivalent to Ω_{train} in all respects, but satisfies $\Omega_{\text{train}} \cap \Omega_{\text{validate}} = \emptyset$. If the MLP gives the same total error for both Ω_{train} and Ω_{validate}, then it is reasonable to expect the MLP to behave as expected when we apply it to Ω_{data}. Hence to ensure that we have not fine tuned the weights of the MLP, we need to check the total error E obtained from the network using Ω_{train}, and then compute the total error E' obtained when the network is applied to Ω_{validate}. When $\partial E / \partial \underline{w}$ has reached a minimum, and both E and $E - E'$ are sufficiently small, we can assume that the weights computed for the MLP are not fine tuned and that the training has converged. Hence we can use the network with confidence on a real data set. In order to determine if E is sufficiently small we have to set an ***error threshold*** δ by hand.

Typically we use either pure reference data samples that resemble the classes we are trying to separate, or Monte Carlo simulated data for Ω_{train} and Ω_{validate}. While the number of events of class A or B used in training can be different, it is generally better to use equal numbers of both types of events in training.

If we reflect upon the large number of weight parameters that have to be determined when we train a neural network, the next logical question is '*How much data do we need to use when training a given network?*'. There has been some discussion on this in the literature; for example, it has been noted that for a MLP with a single hidden layer, with W weight parameters that need determining and an error threshold of δ, then you should use more than W/δ events in the training sample (Baum and Haussler, 1989). For more complicated networks this number is multiplied by a factor of $\ln(N/\delta)$, where N is the number of nodes in the network.

Now we return to the issue of local versus global minima. One can try retraining a neural network, starting with different weight sets to check if the same set of weights is converged upon via the training-validation process. If the same weights are obtained from several different trials, then one has some confidence that the solution obtained may be a global minimum. Another possible test would be to try a different minimisation algorithm, and see if the same solution is obtained.

10.5 Decision trees

The concept of a ***decision tree*** (DT) is derived from that of an optimal cut-based selection of events. As with the previously described methods, the aim of the DT is to separate classes of events with as small a mis-classification error as possible. If one has n dimensions describing classes A and B, then the root node of a DT uses the optimal set of dimensions required to separate A and B with some cut on $\underline{x}_i$. In general, the resulting sub-samples of events will contain both classes, so it is possible to consider further subdivision using an optimal combination of dimensions. This iterative process can be continued until such time as one is able to classify A and B with a satisfactory error rate. Figure 10.8 shows a schematic of a DT. Each of the nodes in the tree will split the data set into an A-like and

Root node

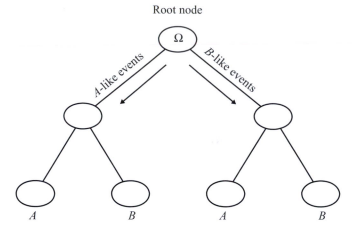

Figure 10.8 A DT with a root node above two layers of nodes that further sub-divide the data sample into pockets of *A*- and *B*-like events.

a *B*-like part. As a result, the lowest level of the tree will contain a number of *A*-like and *B*-like parts. In other words, this layer contains subsets of Ω that are either mostly *A* or mostly *B*, each with a small mis-classification error. As the optimal dimensions are used at each step to separate *A* and *B* it is quite possible that some dimensions will be used more than once while others are never used to classify events. The decision process at each node is equivalent to that of a cut based algorithm. The additional flexibility of a DT of many nodes compared to a cut based optimisation means that the algorithm has more flexibility (hence power) to separate *A* from *B*. The output of a single DT is a binary separation between signal (1) and background (0).

The algorithm used used to separate data into *A*- and *B*-like parts of the data is discussed in Section 10.1, where the number of dimensions used for a given node is that required to provide optimal separation (i.e. not all dimensions have to be used to make the decisions at all of the branching points in a tree). As a result there are between 1 and *n* weight parameters to determine per node in the tree. A corollary of this is that a decision tree with *m* nodes will have between *m* and $n \times m$ weight parameters to determine. While the weight parameter scaling issues of DTs are not as severe as those for a neural network, it follows that the issues discussed above with regard to training validation of weights for neural networks are also serious issues for DTs. Two techniques that can be used to improve the stability of the trained DTs, with respect to statistical fluctuations in the training sample, are boosting (Section 10.5.1) and bagging (Section 10.5.2). In general to avoid over-training a DT one can compute an ensemble of trees with minor variations between them (sometimes referred to as a forest), and use the average classification of an event. In this way one can construct a distribution of outputs with value in the range [0, 1].

10.5.1 Boosting

The aim of training a DT with a ***boosting*** algorithm (referred to as a boosted DT or BDT) is to successively re-weight events in favour of those that are mis-classified from one iteration of the training process to the next. The logic is that the subsequent training iterations will be focused on correctly classifying those events that were previously mis-classified. When boosting a DT one typically re-weights events with a factor α, defined in terms of the error rate ϵ. Having re-weighted events, the total weight of the data sample is renormalised so that the sum of event weights used is constant for all iterations. A number of possible re-weighting factors exist, one of these variants is

$$\alpha = \log\left(\frac{1-\epsilon}{\epsilon}\right), \tag{10.38}$$

which is referred to as the AdaBoost.M1 algorithm, and is discussed at length in Hastie, Tibshirani and Friedman (2009).

10.5.2 Bagging

Bagging is an alternate (or additional) method for improving the stability of a DT to that of boosting. This method involves sampling subsets of data from Ω, and then performing many different training cycles for the DT one for each sub-sample of data. The ultimate set of weights used will be the mean value obtained for the ensemble of DTs. If the data sample Ω is not sufficient to provide statistically distinct subsets of data one can oversample Ω, and use each event many times for different training cycles. This re-sampling method reduces the susceptibility of a DT to statistical fluctuations.

10.6 Choosing an MVA technique

There are a number of factors that should be considered before choosing a particular MVA to separate between classes of events. Some of these factors are logical and based on taking the best classifier to do the job; other factors are subjective and are based on the understanding of the analyst, or indeed the use case of the MVA. If a classifier will be used to provide an end decision on how probable it is that a given element is of class A or B, then your decision to use that classifier might differ

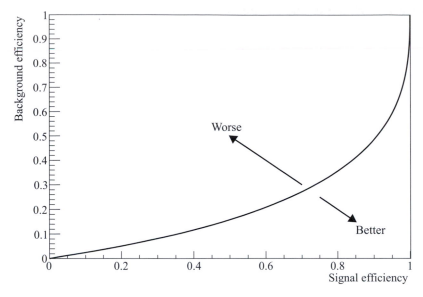

Figure 10.9 The distribution of signal versus background efficiency for a classifier output from an experiment. The better the event classification, the closer the curve will pass to the bottom right-hand corner.

from that made if you intended to use the classifier in a fit based minimisation problem, or indeed as an input to another MVA algorithm.

When assessing the logical input required to understand what is the best classifier, we want to understand how well class *A* is separated from class *B* in our data. There are number of ways to do this; however, it can often be instructive to compute curve of the efficiency of class *A* vs class *B*. In this case we would consider the best classifier to be the one that has the maximum efficiency of one class while minimising the efficiency of the other.

Consider the example from an experiment where we have signal and background classes for *A* and *B*. There are many input variables that distinguish between the classes. These variables are the *n* dimensions that will be used to classify the data. The single output variable from a classifier is then the quantitative information that we have to decide if one algorithm is better than another at separating signal from background. Figure 10.9 shows the distribution of signal versus background efficiency for these hypothetical test data. The better the event classification, the closer it will pass to the bottom right-hand corner. An extreme example of this is the case of being able to identify a sample of pure signal, where the curve will pass through the point $(1, 0)$. Often the rejection rate versus error rate, is plotted when choosing which MVA to use for a problem; such a distribution is known as a receiver operating characteristic.

One can compute a measure of the separation between types A and B by adopting the definition of separation used for the Fisher discriminant in Eq. (10.14) and computing the ratio of the difference of the means (M_i) of the two output distributions and the variances (Σ_i^2),

$$S = \frac{(M_A - M_B)^2}{\Sigma_A^2 + \Sigma_B^2}. \tag{10.39}$$

The greater the value of S for a given classification algorithm, the more discriminating power that has. If the distributions of one or both of the event types as classified by an algorithm are rather different to that of a Gaussian, then this definition of separation may not be very appropriate.

The fact that an algorithm gives optimal separation for the specified set of discriminating variables used, does not necessarily mean that this is the most optimal solution to the problem. In order to ascertain this we would have to compare classifiers for all possible combinations of input discriminating variables and all possible classifier algorithms. Where it is impractical to perform tests with all of the combinations, one should endeavour to test as many as is reasonable before converging upon a candidate classifier to use in the analysis of data. Having identified such a candidate, the next step in the process is to consider any subjective factors as discussed in the following that may influence the decision of an algorithm or number of dimensions to use to classify the data.

When it comes to addressing the subjective input required to understand which classifier is the best, we should remember that there is no magic recipe to help us. However, there are a number of factors that should always be adhered to.

(1) Simplicity can be a key factor in determining which classifier is used. The clarity in understanding what is happening to your data in order for a given event to be classified as one type or another, or indeed to be able to easily explain what you are doing to a colleague should not be underestimated.

(2) Only use a method that you understand. If you use algorithms that you do not fully understand, you may find that you have a better separation between classes; however, you run a risk of having over-trained the algorithm without realising it, or falling foul of some pathological behaviour. The only way to limit such a risk to acceptable levels is to adhere to an abstinence policy of only using algorithms that you understand.

(3) Where appropriate, always ensure that sufficient data are used to train and validate an algorithm. There is a necessary trade off between the desire to use as much data as possible as an input to an algorithm, and ensuring that you can validate that the resulting classifier does not suffer from over-training or some other pathology when checked against a statistically independent

sample. If you fail to validate a classifier that requires training, then you should not be tempted to use that classifier for anything beyond an educational exercise.

(4) Think carefully about the shape of the output classifier in the context of how you wish to use it. For example, consider the following if you are using this as an input to a fit based optimisation.

- Are you able to easily parameterise the target shapes of the classifier?
- If the classifier is a highly irregular, or peaked shape, is there a 1 : 1 mapping that you can apply in order to retain the separation power of the classifier, but obtain a distribution that can be parameterised or used as an input variable in a fit?

(5) Systematic uncertainties resulting from the use of the MVA should be considered. How this is done will depend on how the MVA will be used.

The quantitative and subjective inputs discussed above all play a role when we want to understand which classifier is the best for solving our problem. The discussion in this section is relevant for any MVA technique, not just the algorithms that have been described in detail here. Each problem that you are faced with will have its own unique set of quantitative and subjective factors that must be considered in order to choose which classifier is the best for a given problem. If in doubt, there is little lost in opting for the simplest algorithm. The cost in doing so is usually some loss of precision in a measurement; however, sometimes this can be considered acceptable if the gain in clarity is a subjective factor that carries significant weight for your particular problem.

10.7 Case studies

This section discusses three case studies that can be understood by applying some of the techniques introduced in this chapter.

10.7.1 Determine the cuts

Consider the situation where one expects a uniform background, with a peaking signal described by some Gaussian distribution with a mean of zero and width of one. In this scenario both signal and background event yields are the same, so $N_{signal} \simeq N_{background}$. This scenario can be found in a variety of areas of science. One example would be the measured energy spectrum of X-rays emitted from a decaying nucleus, where the flat background would correspond to noise, and the finite width of the peaking signal would correspond to the intrinsic resolution of the detector being used. In order to maximise the signal to background ratio S/B

one would like to remove as much of the background as possible. Clearly applying a cut on only one side of the distribution will not find an optimal solution as this will have little effect on removing background that is otherwise signal free on the other side of the peak. The solution to this problem is to simultaneously optimise cut positions on both sides of the signal peak. In algorithmic terms this is similar to a two-dimensional optimisation problem with the constraint that the cut value on the low side of the peak can never exceed that on the upper side of the peak.

In practice one can again perform a numerical integration in order to obtain the optimal set of cuts; however, for small samples of (simulated) data, one may be be susceptible to fluctuations in the cut values about the true optimal solution. Given that in this case there are well-defined expectations for the signal and background distributions, one could solve the problem analytically if so desired.

10.7.2 *SPAM filtering*

Many e-mails in circulation are SPAM, and so rather than wade through such mail in order to identify those of interest, it is desirable to be able to automatically identify mails that could potentially be considered SPAM and have them moved to a separate mail folder for inspection prior to deletion. This problem is commonplace, and the issue of SPAM filtering using a Bayesian algorithm with inputs from both text and domain information was first raised by Sahami, Dumais, Heckerman and Horvitz (1998).

Here the issue is simple: We want to be able to retain all of our legitimate e-mails to read, without having to select and delete every single SPAM e-mail. One thing that is worth noting is that the penalty of making a type I error and classifying a legitimate e-mail as a SPAM mail is far greater than the penalty of making a type II error and classifying a SPAM mail as a legitimate mail. One can use a Bayesian classifier as a SPAM filter. The filter considered in the following is simpler than that discussed in the reference.

An event in this context corresponds to an e-mail, including the header information, the subject, and text body. Each unique element of information in an e-mail corresponds to a dimension of the problem space, and one can assign the frequency of information occurring as a value for that particular dimension. Thus one has an n-dimensional vector $\underline{x}$. For example a simple e-mail greeting may include the salutation 'Dear Mr(s) Smith', therefore the dimension associated with the word 'Dear' would have a value of one, and so on. Having re-arranged an e-mail into a vector of information $\omega_i(\underline{x})$, the next step is to compute $P(\omega_i(\underline{x})|LEGITIMATE)$ and $P(\omega_i(\underline{x})|SPAM)$, or some ratio as discussed in Section 8.4. In order for this to happen one has to assign a significance of a given dimension in the computation of the probability that the event is legitimate or SPAM. A possible metric to use for this is

$$P(\omega_i|H_i) = \prod_{k=1}^{m} P(x_{ik}|H_i), \qquad (10.40)$$

where x_{ik} is the value of the kth dimension for the ith e-mail. In other words, the probability of an e-mail being legitimate (or SPAM) is given by the product of the probabilities of all elements of the e-mail being consistent with a legitimate (or SPAM) mail. In making this assertion one is assuming that all elements of the e-mail are independent of each other, so for example the individual words in the phrase 'Dear Mr(s) Smith' are un-correlated. In order to determine the $P(x_{ik}|H_i)$ for a given word or combination of words, one needs to have access to a sample of training data, where the target type is know with certainty. Given this it is possible to compute $P(x_{ik}|H_i)$, hence the effectiveness of a Bayesian classifier to correctly classify legitimate and SPAM e-mail. As mentioned at the start of this section, this example is a simpler method than that proposed by Sahami, Dumais, Heckerman and Horvitz (1998). In fact the classification of text-based information is an active field of research with very practical ramifications that goes beyond the use of Bayesian classifiers. The interested reader might wish to consult Srivastava and Sahami (2009) for an overview of this topic.

10.7.3 MVA selection problem

The use of multivariate methods in particle physics experiments has gained in popularity since the advent of modern computing, and the development of centralised tools to facilitate training and validation of the underlying algorithms. Modern measurements often attempt to extract information from signal events that are dominated by background; for example, it is not unusual to study samples of a few hundred signals in a selected sample of 50 000 or 100 000 events. Each time one constructs such an analysis, it is necessary to understand what an optimal choice of MVA is for a given set of algorithms and a given set of input variables. Figure 10.10 shows the Fisher, MLP and BDT distributions obtained for signal Monte Carlo and background data control samples computed using data from the BABAR experiment. These discriminating variables combine eight input variables that provide information on the distribution of energy recorded in the detector systems of BABAR. One can see statistical fluctuations in these shapes as only 1000 signal and background events were used for training. The corresponding receiver operating characteristic is also shown in the figure. In general the MLP provides the best separation of

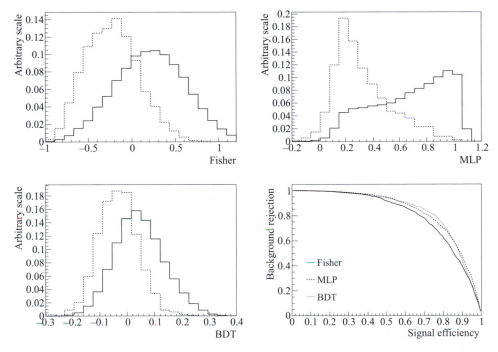

Figure 10.10 Example distributions obtained for (solid) signal and (dashed) background training samples for (top left) Fisher discriminant, (top right) MLP, and (bottom left) BDT algorithms. The corresponding receiver operating characteristic is also shown (bottom right).

signal and background in high-purity regions; however, the BDT provides a better separation than the MLP in the overlap region. If one of these MVA distributions is incorporated in a fit to data then it may be important to consider if the more Gaussian-like shape of the BDT or Fisher provides an advantage over the MLP distribution.

10.8 Summary

The main points introduced in this chapter are listed below.

(1) It is often possible to combine information from a multi-dimensional space of discriminating variables into a single output that can be used to separate two or more classes of data. There are many algorithms described in the literature that can address this problem, each with their own benefits and limitations.

(2) The algorithms discussed here are as follows:

- cut-based selection (Section 10.1),
- Bayesian classifier (Section 10.2),
- the Fisher linear discriminant (Section 10.3),
- artificial neural networks (Section 10.4),
- decision trees, DT (Section 10.5).

(3) Classification algorithms that require an iterative training process must be validated in some way. Failure to validate training could result in the inappropriate use of a non-optimal classifier. Section 10.4.3 discussed training, and validation techniques are reviewed in Section 10.4.4.

(4) There is no right or wrong solution to the question of which classifier to use in a particular problem. If in doubt, the simplest solution to the problem that is well behaved and understood may be adequate.

(5) In general, the most optimal solution to a problem should be adopted to classify events. However, if the user does not understand an algorithm, they should be discouraged from applying the results of a 'black box' to their problem for the simple case that there may be pathologies in that solution which have been overlooked.

Exercises

10.1 Compute the coefficients of a Fisher discriminant to separate samples of data A and B given the means μ and covariance matrices σ^2 below:

$$\mu_A = \begin{pmatrix} 1.0 \\ 2.0 \end{pmatrix}, \quad \mu_B = \begin{pmatrix} 0.0 \\ 1.0 \end{pmatrix}, \quad \sigma_A^2 = \begin{pmatrix} 1 & 2 \\ 2 & 2 \end{pmatrix}, \quad \sigma_B^2 = \begin{pmatrix} 1 & 1 \\ 1 & 2 \end{pmatrix}.$$

10.2 Compute the coefficients of a Fisher discriminant to separate samples of data A and B given the means μ and covariance matrices σ^2 below:

$$\mu_A = \begin{pmatrix} 0.0 \\ 1.0 \end{pmatrix}, \quad \mu_B = \begin{pmatrix} -1.0 \\ 2.0 \end{pmatrix}, \quad \sigma_A^2 = \begin{pmatrix} 2 & 1 \\ 1 & 1 \end{pmatrix}, \quad \sigma_B^2 = \begin{pmatrix} 1 & 0 \\ 0 & 2 \end{pmatrix}.$$

10.3 Compute the coefficients of a Fisher discriminant to separate samples of data A and B given the means μ and covariance matrices σ^2 below:

$$\mu_A = \begin{pmatrix} 0.2 \\ 0.7 \\ 0.1 \end{pmatrix}, \qquad \mu_B = \begin{pmatrix} 0.6 \\ 1.5 \\ 0.2 \end{pmatrix},$$

$$\sigma_A^2 = \begin{pmatrix} 0.1 & 0.0 & 0.0 \\ 0.0 & 0.1 & 0.0 \\ 0.0 & 0.0 & 0.2 \end{pmatrix}, \quad \sigma_B^2 = \begin{pmatrix} 0.15 & 0.0 & 0.0 \\ 0.0 & 0.3 & 0.0 \\ 0.0 & 0.0 & 0.3 \end{pmatrix}.$$

Table 10.1 *Data corresponding to* (x, y)
points for classification.

i	1	2	3	4	5	6
x_i	-1.0	0.0	1.0	3.0	2.0	0.5
y_i	1.0	2.0	1.0	3.0	-1.0	1.0

10.4 Compute the coefficients of a Fisher discriminant to separate samples of data A and B given the means μ and covariance matrices σ^2 below:

$$\mu_A = \begin{pmatrix} 1.0 \\ 0.7 \\ 0.5 \end{pmatrix}, \qquad \mu_B = \begin{pmatrix} 0.6 \\ 1.0 \\ 0.2 \end{pmatrix},$$

$$\sigma_A^2 = \begin{pmatrix} 0.2 & 0.1 & 0.0 \\ 0.1 & 0.1 & 0.0 \\ 0.0 & 0.0 & 0.2 \end{pmatrix}, \qquad \sigma_B^2 = \begin{pmatrix} 0.15 & 0.0 & 0.1 \\ 0.0 & 0.3 & 0.0 \\ 0.1 & 0.0 & 0.3 \end{pmatrix}.$$

10.5 The ratio of energy and momentum (E/p) for a charged particle can be used to distinguish between different species (for example e, μ, and π) and has a value between zero and one. Define a Bayesian classifier based on Gaussian distributions with $\mu = 1.0$ for an electron, 0.5 for a π meson and 0.0 for a μ, all three PDFs having a width $\sigma = 0.2$, and classify the following events: $E/p = 0.0, 0.1, 0.2, 0.3, 0.4, 0.5, 0.6, 0.7, 0.8, 0.9, 1.0$.

10.6 A data sample Ω of events is composed of signal and background components. The signal is distributed according to a Gaussian PDF with $\mu = 0.0$ and $\sigma = 1.0$, whereas the background is uniform. Use a Bayesian classifier to categorise the events: $\Omega = \{-3, -2, 0, 1, 4, 5\}$.

10.7 Use a Bayesian classifier to categorise events for the following problem. Two signals are distributed in an (x, y) space, with two-dimensional unit Gaussian distributions centred on $(0, 0)$ and $(1, 2)$, for types A and B, respectively. Classify the events in Table 10.1 according to this model.

10.8 Use a Bayesian classifier to categorise events from Table 10.1 using the following model: type A are distributed uniformly, and type B are distributed according to a two-dimensional unit Gaussian centred at $(1, 2)$.

10.9 Signal events are distributed according to $P_A = 3(2x - x^2)/4$, and background events are uniform ($P_B = 1/2$) in the interval $x \in [0, 2]$. It is expected that there are 10 signal and 20 background events in the data.

Optimise the cut value X_c with regard to the test statistic signal/background in the interval $[0, X_c]$.

10.10 Do the values for the signal and background yields given in the previous exercise play a role in the optimal cut value obtained?

10.11 Compute the separation between two distributions of events where $\mu_A = 1$, $\sigma_A = 2$ and $\mu_B = 3$, $\sigma_B = 1$.

10.12 Compute the separation between two distributions of events where $\mu_A = 1$, $\sigma_A = 1$ and $\mu_B = 0$, $\sigma_B = 1$.

10.13 Signal events are distributed according to $P_A = 6(x - x^2)$, and background events are uniform ($P_B = 1$) in the interval $x \in [0, 1]$. Optimise the cut value X_c with regard to the test statistic signal/background in the interval $[0, X_c]$.

10.14 Signal events are distributed according to $P_A = 12(x^2 - x^3)$, and background events are uniform ($P_B = 1$) in the interval $x \in [0, 1]$. Optimise the cut value X_c with regard to the test statistic signal/background in the interval $[0, X_c]$.

10.15 How can one improve the signal to background significance found in the previous exercises?

Appendix A

Glossary

This section summarises many of the terms used in this book. A brief description of the meaning of each term is given, and where appropriate a cross reference to the corresponding section in the text is suggested for further reading.

$\wedge$: Logical and.
$\vee$: Logical or.
χ^2: See chi-square.
μ, $\mu(i)$ or μ_i: See mean.
ρ, $\rho(x, y)$ or ρ_{xy}: See correlation.
σ, $\sigma(i)$ or σ_i: See standard deviation.
σ_{xy} or $\sigma(x, y)$ or σ_{ij}: See covariance.

Alternate hypothesis (H_1): The complement of a null hypothesis. If a statistical test is performed for a null hypothesis, and that test fails, then the alternate hypothesis (as the complement of the null hypothesis) is the result supported by the data.

Arithmetic mean: See mean.

Artificial neural network: Also called a neural network. This is a collection of perceptrons fed by some number of input variables, with one or more output nodes. The function of the neural network is to try and optimally distinguish between two or more types of data. See Section 10.4.

Bagging: This is the name given to a technique of oversampling data used for training decision trees. See Section 10.5.2.

Bayes' theorem: See Eq. (3.4).

Bayesian classifier: A classification algorithm based on Bayes' theorem, see Section 10.2.

Bayesian statistics: The branch of statistics derived from the work of Rev. Bayes. This is a subjective approach to the computation of probability that can be used in a variety of situations, including for those that are not repeatable. See Section 3.2.

Binomial distribution: A distribution that describes a bi-modal situation, and is often presented in terms of success with some probability p, and failure with some probability $q = 1 - p$. See Section 5.2.

Binomial error: The error on an observable that is bi-modal, such as the detection efficiency, is given by a binomial error. See Section 6.3.

Boosting: This is the name given to a re-weighting of events used for training decision trees. See Section 10.5.1.

Blind analysis: A method of validating and performing an analysis while not being able to determine the true value of the measured observable until the last moment in the measurement process. This methodology can be used to try and minimise experimenter bias, although it is not applicable in all circumstances. Blind analysis is discussed in more detail in Section 6.6.

Breit–Wigner: A probability density function often used to describe resonant behaviour in physics. See Appendix B.1.2.

Chi-square: The sum of squared deviations of data x_i relative to some model $\theta(x_i)$ normalised by the standard deviation of the ith point. See Section 5.5. Also used in the context of a fit; see Chapter 9.

Combination of errors: The process of computing the error on some function f depending on an number of measurements of observables $x \pm \sigma_x$, $y \pm \sigma_y$, etc., as discussed in Section 6.2.

Confidence level: The confidence level of $X\%$ is a statement about the value of some observable obtained as the conclusion of a measurement. Assuming that the original measurement corresponds to the true value, a repeat measurement would yield the same or a compatible conclusion $X\%$ of the time. The corollary of this is that $100 - X\%$ of the time, one can obtain an equally valid result that would be in disagreement with the original conclusion.

Confidence interval: The interval in some variable corresponding to a given confidence level.

Correlation: The correlation between two variables x and y is is a measure of how dependent they are on each other. The symbol often used to denote correlation is ρ (with or without subscripts). If the correlation ρ_{xy} is zero, then x and y are independent, and if the correlation has a magnitude of one, then having determined the value one of the variables, the other can be determined as well. See Section 4.6.2.

Correlation matrix: For two or more variables, it is possible to construct a matrix of correlation coefficients, referred to as the correlation matrix.

Covariance: The covariance of two variables x and y is the term given to the average of the sum of the products of residuals of the variables x and y relative to the mean value of those variables. The covariance between x and y is denoted by σ_{xy}, and it is related to the correlation ρ_{xy} by $\sigma_{xy} = \rho_{xy}\sigma_x\sigma_y$, where the sigmas refer to the standard deviations of x and y. See Section 4.6.1.

Covariance matrix: For two or more variables, it is possible to construct a matrix of variance and covariance coefficients, referred to as the covariance matrix.

Coverage: Corresponds to the confidence level with which an interval or limit is defined by. Under coverage refers to the situation where a limit or interval with a quoted CL actually has less coverage than stated. Over coverage refers to the opposite scenario where the quoted limit or interval has more coverage than stated. Both under coverage and over coverage are undesirable.

Cut-based selection: A multivariate technique to distinguish between types or classes of data. See Section 10.1.

Decision tree: A multivariate technique used to distinguish between types or classes of data. See Section 10.5.

Error: There are several types of error that one might consider.

- A mistake.
- See type I error.
- See type II error.
- See standard error on the mean (or standard deviation).

- See combination of errors.
- Event mis-classification in a multivariate algorithm.

The context that the term error is used in should make the meaning of the word clear.

Error matrix: See covariance matrix.

Expectation value: The mean result expected from a given distribution or measurement.

Extended maximum likelihood fit: The use of an extended likelihood function to fit to a sample of data. See Section 9.4.1.

False negative: See type I error.

False positive: See type II error.

Fisher discriminant: A multivariate technique used to distinguish between types or classes of data. See Section 10.3.

Fit/fitting: Given a set of data $\underline{D}$, the process of optimising a set of parameters $\underline{p}$ of some model θ in order that the model describes the data in the best possible way is called fitting. One fits the model to the data.

Frequentist statistics: The branch of statistics derived from the axiom that one can repeat measurements many times. A frequentist probability is determined from taking the limit that the number of repeated measurements tends to infinity. Often it is possible to exactly compute a frequentist probability without taking this limit. This method can not be used if a measurement is not repeatable. See Section 3.4.

Full width at half maximum (FWHM): The full width at half maximum (FWHM) of a function $f(x)$ is the distance in x between the two points on $f(x)$ that are at half of the global maximum of the function. It is usually only sensible to use the FWHM for functions that have a single global maximum such as a Gaussian or Breit–Wigner function.

Gaussian distribution: A probability density function often used to describe measurements made using large samples of data. See Section 5.4.

Goodness of fit: A quantitative measure of how well a data and model agree with each other. See Section 5.5.

Gradient decent: A numerical method used for optimisation of some test statistic. See Section 9.1.1.

Graph: A visual display representing a set of points or a function in two or three dimensions. See Section 4.1.2.

Histogram: A binned graphical representation of data. See Section 4.1.1.

Hypothesis: A postulate compared against a sample of data.

Least-squares fit: A fitting technique derived from the χ^2 distribution for a set of data. See Section 9.3, also see chi-square.

Likelihood: The likelihood function is a probability density function used to model a sample of data. See Section 3.6.

Maximum likelihood fit: The use of a likelihood function to fit to a sample of data. See Section 9.4.

Mean: A measure of the central value of a distribution. See Section 4.2. The mean is also called the arithmetic mean or average.

Median: A measure of the central value of a distribution. See Section 4.2. The median of some distribution in a variable x is the value of x that corresponds to the midpoint in the distribution where there are equal numbers of data both above and below that point.

Mode: A measure of the central value of a distribution. See Section 4.2. The mode of some distribution is the most frequently occurring value. If the data in a distribution are binned as a histogram, then the mode will correspond to the bin that contains the largest number of entries.

Monte Carlo: Monte Carlo is the term given to a technique of using sequences of random numbers in order to simulate or model a given situation.

Multi-layer perceptron: A type of artificial neural network based on more than one layer of perceptrons, with several input variables, and one or more output variable. See Section 10.4.2.

Null hypothesis (H_0): The default hypothesis used to test against a sample of data. Such a test will attempt to ascertain if the null hypothesis is compatible with the data.

Optimisation: The process of varying the parameters of some model or algorithm in order to minimise or maximise some test statistic. See Section 9.1.

***p*-value**: This is the probability for obtaining a result corresponding to that observed in data, or a more extreme fluctuation.

Perceptron: A perceptron is a logical unit that is used to process a set of input values, and compare against some threshold function in order to determine a given output. See Section 10.4.1.

Poisson distribution: The Poisson distribution is a probability density function used to describe processes where there is no way of knowing how many events have occurred. This distribution has the property that the mean and variance are the same value. See Section 5.3.

Probability: A measure of the chance of something happening. This concept is discussed in Chapter 3.

Probability density function (PDF): A function used to describe a distribution of some data. The total integral over the interesting problem space of a PDF is one. See Section 3.5. A number of useful PDFs are described in Chapter 5 and Appendix B.

Rectangle rule: A numerical integration algorithm (see Appendix C.1).

Relative error: The error on a quantity expressed as a fraction or percentage of the value of the quantity.

Set: A collection of unique elements. See Chapter 2.

Skew: The skew of a distribution is a measure of how asymmetric a distribution is about the mean value. A symmetric distribution has a skew of zero. See Section 4.5.

Standard deviation: A measure of the spread of data in some distribution, denoted by the symbol σ. See Section 4.3.2.

Standard error on the mean: See standard deviation.

Statistic: A term ascribed to mean some calculable quantity. For example arithmetic mean, variance, χ^2 are statistics. Note that often physicists refer to the number of data using the same term, however the meaning inferred should be clear from the context.

Statistical error: The error determined from a measurement of some observable, whose sources are purely the random result performing the measurement. Also known as the statistical uncertainty.

Systematic error: An error that is the result of some assumption explicitly or implicitly made in a measurement that may arise in the central value measured being different from the that of a perfect measurement using the same data.

Training: The process of determining a set of coefficients used in a multivariate technique in order to distinguish between classes or types of event.

Training validation: The process of validating the training procedure, and *optimal* coefficients obtained for a multivariate technique.

Trapezium rule: A numerical integration algorithm (see Appendix C.2).

Type I error: Rejecting the null hypothesis in a test, where in fact the null hypothesis provides the correct conclusion. Thus a type I error is a false negative response.

Type II error: Accepting the null hypothesis in a test, where in fact the null hypothesis does not provide the correct conclusion. Thus a type II error is a false positive response.

Uncertainty: This is the spread with which it is possible to determine the value of a parameter via some measurement.

Upper limit: If on performing a search for some effect, one fails to obtain a definite indication that an effect may or may not exist, then one can ascribe a numerical bound on the non-existence of the effect. Above this bound, the effect will have been ruled out, and below this bound, the effect may still exist. This bound is called an upper limit, and as with a confidence interval, there is an associated confidence level ascribed to an upper limit. See Section 7.2.

Variance: The square of the standard deviation.

Venn diagram: A graphical way of describing sets. See Chapter 2.

Appendix B

Probability density functions

This appendix summarises the functional forms of some commonly used probability density functions (PDFs) and follows on from Chapter 5. In the following the functional form of the distribution is introduced, and is accompanied with a graphical representation of the PDFs. Several of the PDFs described are specific to problems relating to physics, such as resonant behaviour, and specialist particle physics applications. Where appropriate, there is some elaboration of the physically relevant PDFs for completeness. If there are numerical precision issues relating to the computation of a particular shape, or determination of one or more parameters when used in an optimisation procedure this is noted. Similarly some of the functional forms quoted are not normalised to unity over all space.

While the functional forms discussed here can be used to construct complicated models to represent or test data, sometimes it is necessary to combine PDFs in order to use a composite PDF that is the sum of several parts. This can be achieved by adding together different functional forms with an appropriate weighting factor for each PDF in order to retain a total probability of unity, i.e.:

$$\mathcal{P} = f_1 \mathcal{P}_1 + f_2 \mathcal{P}_2 + \cdots \left(1 - \sum_{i=1}^{n-1} f_i \right) \mathcal{P}_n, \tag{B.1}$$

where the sum of the coefficients is unity. If each of the $\mathcal{P}_i$ are normalised, then the total PDF $\mathcal{P}$ will be properly normalised. If this is not the case, then one has to analytically or numerically integrate $\mathcal{P}$ over the domain of interest to obtain an appropriate normalisation constant.

B.1 Parametric PDFs

Most PDFs used in modelling physical situations are parametric PDFs. Such PDFs depend on one or more discriminating variables represented by either a scalar x or vector $\underline{x}$, and the specific shape of the PDF can be modified by changing one or more parameters $\underline{p}$. In general terms we can write a PDF that depends on several discriminating variables, and parameters as $\mathcal{P}(\underline{x}; \underline{p})$. In order for the function to represent a PDF, it must satisfy the normalisation condition

$$\int_{\underline{A}}^{\underline{B}} \mathcal{P}(\underline{x}; \underline{p}) d\underline{x} = 1, \tag{B.2}$$

where the limits $\underline{A}$ and $\underline{B}$ represent the physically interesting space of the discriminating variables that one wants to use. It follows that, for a one-dimensional PDF, we have

$$\int_A^B \mathcal{P}(x; \underline{p}) dx = 1, \tag{B.3}$$

where in general we assume that the PDF may depend on several parameters. If a PDF is not naturally normalised to conserve probability, then one has to determine the corresponding normalisation constant required in order to maintain the above relationship. If it is possible to analytically determine a normalisation constant, then it is advisable to do so as this will generally be more efficient (in terms of computing resources) than performing an accurate numerical integration of a function (see Appendix C).

The systematic uncertainty associated with the use of a parametric PDF when extracting some quantity from data comes from both the lack of knowledge of parameters, as well as the dependence on the choice of the PDF itself. If the chosen PDF agrees well with the data, then it is likely that using another very similar PDF will produce essentially the same result, and that this second systematic uncertainty is small, or negligible. Uncertainties arising from the lack of knowledge can be addressed by computing the deviation of a result obtained when varying some parameter p_i by its uncertainty, $\pm 1\sigma_{p_i}$. For uncorrelated parameters, the sum in quadrature of the σ_{p_i}s provides the systematic uncertainty from this source. This approach overestimates the systematic uncertainty for correlated parameters.

B.1.1 Binomial

The binomial distribution (see Section 5.2) for n trials is given by

$$P(r; p, n) = p^r (1-p)^{n-r} \frac{n!}{r!(n-r)!}, \tag{B.4}$$

where r trials are successful and the probability of success is given by p.

B.1.2 Breit–Wigner

The **Breit–Wigner** distribution (also called a non-relativistic Breit–Wigner or Cauchy function) is given by

$$\mathcal{P}(x; m_0, \Gamma) = \frac{1}{\pi} \frac{\Gamma}{(x - m_0)^2 + (\Gamma)^2} \tag{B.5}$$

and is shown in Figure B.1. Here m_0 is the position of the peak and Γ is the width of the peak. By construction the distribution is symmetric about the mean value m_0, with a characteristic width given by Γ. This is a properly normalised PDF where

$$\int_{-\infty}^{+\infty} \mathcal{P}(x; m_0, \Gamma) dx = 1. \tag{B.6}$$

The Breit–Wigner distribution is often used to describe resonant behaviour in nature, for example that exhibited in electronic circuits such as the LCR circuit, and in particle or nuclear interactions where scattering of particles will be enhanced at a resonance.[1] A

---

[1] Although it should be noted that modified forms of the Breit–Wigner exist to describe resonances where relativistic factors become important, or where non-zero spin quantum numbers describe the underlying dynamics,

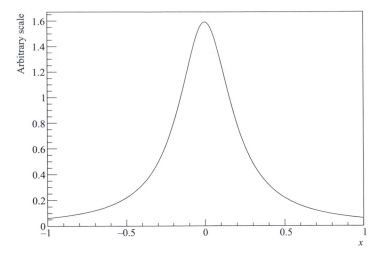

Figure B.1 An example of the Breit–Wigner (Cauchy) PDF.

detailed introduction of the physical significance of this PDF can be found elsewhere, for example in the book (Feynman, Leighton and Sands, 1989).

B.1.3 χ^2

This distribution is defined by χ^2 and the number of degrees of freedom ν, which is given by the number of data N less the number of constraints c imposed. The minimum value for c is one, which is the constraint arising from the total number of data being used in the comparison. The χ^2 distribution has the following form

$$P(\chi^2, \nu) = \frac{2^{-\nu/2}}{\Gamma(\nu/2)}(\chi^2)^{(\nu/2-1)}e^{-\chi^2/2}, \tag{B.7}$$

where

$$\Gamma(\nu/2) = (\nu/2 - 1)!, \tag{B.8}$$

for positive integer values of ν, and $\Gamma(1/2) = \sqrt{\pi}$. This distribution is described in Section 5.5.

B.1.4 Exponential

The **exponential distribution** defined as

$$\mathcal{P}(x; \gamma) = \frac{1}{N}e^{\gamma x}, \tag{B.9}$$

where γ is the slope (or constant) of the exponential (see Figure B.2). Often when dealing with the behaviour of physical processes as a function of time (or some other variable) it can be useful to replace the constant γ with $1/\tau$, where the parameter τ is the characteristic

for example see the Dalitz plot analysis formalism review compiled by the Particle Data Group (Beringer *et al.*, 2012). In general relativistic Breit–Wigner distributions are asymmetric about the modal value.

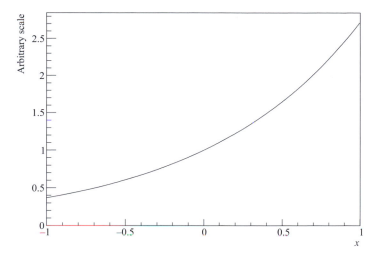

Figure B.2 An example of the exponential PDF with a positive value of γ.

decay rate (for negative values of γ) with the same units as the variable x. The normalisation constant N is simply given by the integral of the exponential over the desired range in x from A to B, i.e.

$$N = \frac{e^{\gamma B} - e^{\gamma A}}{\gamma}.$$ (B.10)

B.1.5 Gaussian

This is defined by a mean and width (μ and σ) and is given by

$$\mathcal{P}(x\,;\mu,\sigma) = \frac{1}{\sigma\sqrt{2\pi}} \exp\left(-[x-\mu]^2/2\sigma^2\right)$$ (B.11)

where the integral from $-\infty$ to $+\infty$ is one. The Gaussian PDF is described in detail Section 5.4, and the multivariate normal distribution is discussed in Section 7.8.1. An asymmetric Gaussian distribution is a Gaussian distribution where the values of σ are allowed to differ above and below $x = \mu$.

B.1.6 Gaussian with an exponential tail

A Gaussian with an exponential tail can be used to parameterise data that has a core part of the distribution resembling a Gaussian, but one that also has a significant skew. In limited circles this distribution is also known as the 'Crystal Ball' distribution (Oreglia, 1980; Gaiser, 1982; Skwarnicki, 1986). The functional form of this distribution is (see Figure B.3)

$$\mathcal{P}(x;m_0,\sigma,\alpha,n) = \begin{cases} \frac{1}{N} \times e^{-(x-x_0)^2/(2\sigma^2)} & x > x_0 - \alpha\sigma, \\ \frac{1}{N} \times \frac{(n/\alpha)^n \, \exp(-\alpha^2/2)}{((x_0-x)/\sigma + n/\alpha - \alpha)^n} & x \le x_0 - \alpha\sigma. \end{cases}$$ (B.12)

By allowing the parameters α and n to vary when determining the PDF, one will find that these parameters are all highly correlated. If there is some way of determining a suitable

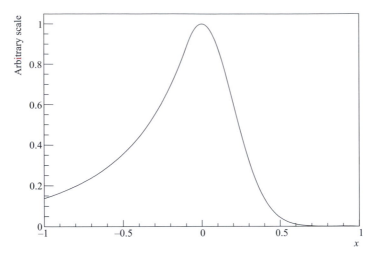

Figure B.3 An example of a Gaussian with an exponential tail (Crystal Ball) PDF.

value for one or both of these parameters using prior information, that will result in a faster convergence of any fit that is performed using this PDF. The normalisation constant N is given by the integral of the PDF over the problem domain in x.

B.1.7 Landau

This is an important PDF, in particular, for nuclear and particle physicists. The functional form of the Landau distribution is given by Landau (1944), and this is used to describe the fluctuations in energy loss of a charged particle passing through a thin layer of material. Numerically this PDF is time consuming to compute, and there are a number of algorithms that have been used over the years in order to evaluate an estimate of the Landau PDF.

Figure B.4 shows an example of the Landau PDF, which is given by

$$\phi(\lambda) = \frac{1}{2\pi i} \int_{\sigma-i\infty}^{\sigma+i\infty} e^{u\ln u + \lambda u} du, \tag{B.13}$$

where there are two parameters describing the distribution, the most probable value given by λ_0 and a width parameter $\sigma \geq 0$. The underlying numerical algorithm used in order to compute the distribution shown in the figure is described by Kölbig and Schorr (1984). A modern discussion of the computation of this function, with references to earlier works can be found in Fanchiotti, Garcia-Canal and Marucho (2006).

B.1.8 Poisson

The Poisson distribution (Section 5.3) for r observed events is given by

$$P(r, \lambda) = \frac{\lambda^r e^{-\lambda}}{r!}, \tag{B.14}$$

where λ is the mean and variance of the distribution.

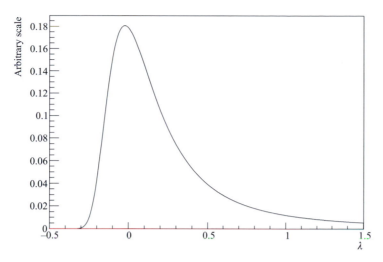

Figure B.4 An example of the Landau PDF with a most probable value of $\lambda_0 = 0.0$, and width parameter $\sigma = 0.1$.

B.1.9 Polynomial

One way to define a ***polynomial distribution*** is given by

$$\mathcal{P}(x; p_k) = \frac{1}{N} \sum_{k=0}^{M} p_k x^k, \tag{B.15}$$

where the p_k are coefficients. The normalisation constant N is given by the integral over the interesting domain $A \leq x \leq B$ for the sum

$$\int_{A}^{B} \mathcal{P}(x; p_k)dx = \left[\sum_{k=1}^{M} k \times p_k x^{k-1} \right]_{A}^{B} = N. \tag{B.16}$$

In general the coefficients p_k will be correlated with each other. In particular the odd terms (odd k) will be strongly correlated with each other, and the even coefficients (even k) will be strongly correlated with each other. For this reason, it may be desirable to limit the order of a polynomial, or for example use only odd (even) terms for odd (even) distributions of events in order to limit the number of parameters allowed to vary when fitting data. Figure B.5 shows an example of a cubic function. It is possible to define a polynomial distribution in other ways, so if you are using a particular representation, ensure that you use this consistently. Furthermore, if you find that the polynomial coefficients p_k are highly correlated with each other for a particular problem, you may find using an alternative definition of a polynomial function, for example one formed from a set of orthogonal terms, to be a better choice of PDF. Examples of orthogonal polynomials include Legendre, Chebychev, and Hermite polynomials.

One final thing to remember; if a polynomial function is considered a suitable PDF that is related to the probability of some event to occur, then this PDF needs to be non-negative over the relevant domain. This follows as probability is a positive definite quantity bound by zero and one.

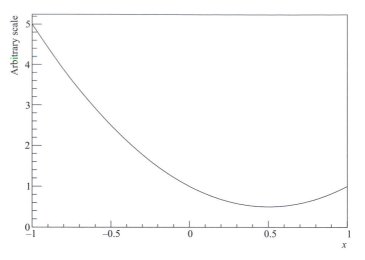

Figure B.5 An example of the polynomial PDF.

B.1.10 Sigmoid

The sigmoid (or logistic) function is often used in machine learning algorithms or to model behaviour at some threshold. The sigmoid function is given by

$$\mathcal{P}(x; a, b) = \frac{1}{N} \frac{1}{1 + e^{ax+b}}, \tag{B.17}$$

where the coefficients a and b are an exponent scale factor and x-offset, respectively. The factor N is a normalisation constant that has to be determined for the problem domain in x from x_1 to x_2. This can be obtained by substituting a variable for the denominator and integrating Eq. (B.17) using partial fractions. The result obtained for the normalisation is

$$N = \left[x - \frac{\ln|1 + e^{ax+b}|}{a} \right]_{x_1}^{x_2}. \tag{B.18}$$

Figure B.6 shows an example of the sigmoid PDF.

B.1.11 Step and veto functions

It can be useful to impose a sharp cut-off to a PDF. In such cases it is useful use a ***step function*** PDF where

$$\mathcal{P}(x) = \begin{cases} \frac{1}{N}, & x_a < x < x_b \\ 0, & x < x_a \text{ or } x > x_b. \end{cases} \tag{B.19}$$

The normalisation N for a step function depends on the values of x_a and x_b, as can be seen from the following:

$$\int_{x_a}^{x_b} \mathcal{P}(x)dx = \int_{x_a}^{x_b} \frac{1}{N} dx = \frac{(x_b - x_a)}{N} = 1. \tag{B.20}$$

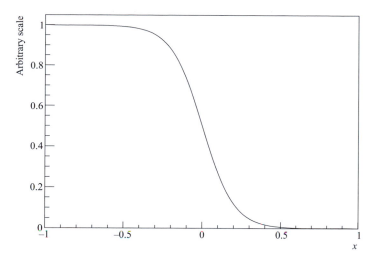

Figure B.6 An example of the sigmoid PDF.

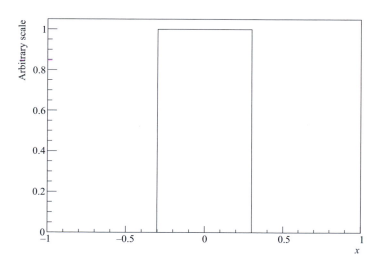

Figure B.7 An example of a step function.

Thus in order to satisfy the normalisation condition that the total probability is one, $N = x_b - x_a$, thus

$$\mathcal{P}(x) = \frac{1}{(x_b - x_a)}, \tag{B.21}$$

Figure B.7 shows an example of the step function where $x_a = -0.3$ and $x_b = 0.3$.

The complement of a step function is a ***veto function***, where the PDF has a value of zero for finite range $x_a < x < x_b$, and is otherwise constant. Thus the functional form of the veto function is

$$\mathcal{P}(x) = \begin{cases} \frac{1}{N}, & x < x_a \text{ or } x > x_b \\ 0, & x_a < x < x_b \end{cases}. \tag{B.22}$$

The normalisation for the veto function depends not only on x_a and x_b, but also on the domain boundaries of interest A and B. Thus

$$\int_A^B \mathcal{P}(x)dx = \int_A^{x_a} \frac{1}{N} dx + \int_{x_b}^B \frac{1}{N} dx \tag{B.23}$$

$$= \frac{x_a - A + B - x_b}{N}, \tag{B.24}$$

hence in order for this PDF to be properly normalised we require $N = x_a - A + B - x_b$.

B.2 Non-parametric PDFs

Those PDFs where there are no parameters p used to determine the shape are called ***non-parametric*** PDFs. For such functions, the PDF is completely pre-determined and can not vary in an optimisation process. Such PDFs can be useful when control sample data show complicated behaviour that would be difficult to parameterise in a sensible way. Several examples of non-parametric PDFs are discussed in the following. The PDF normalisation constants for non-parametric PDFs are generally determined numerically (see Appendix C for a simple introduction to this topic). If the non-parametric PDF is a binned PDF, then the normalisation can be determined by simply summing the weights of each bin in that distribution.

It should be noted that unlike parametric PDFs, where one can compute systematic uncertainties on an extracted quantity from the lack of knowledge of parameters, as well as dependence on the choice of PDF, the systematic uncertainty associated with the use of non-parametric PDFs arises from dependence on the choice of PDF and binning. Care should be taken to address this issue properly when analysing data.

B.2.1 Histogram

It is possible to construct a non-parametric PDF based on an input histogram. Such a PDF will have a pre-determined level as a function of the discriminating variable. The function will take the form

$$\mathcal{P}(x) = B(x), \tag{B.25}$$

where B is a discrete function with a constant value across a bin interval. Figure B.8 shows an example of a histogram PDF.

While the content of a histogram is pre-determined but discrete, it may be considered undesirable to have a discrete distribution as a PDF. One can interpolate between the bin content of adjacent bins to determine a more smoothly varying form for $\mathcal{P}$ at the bin boundaries. Such a process is referred to as smoothing a histogram. In practice this finesse usually results in an improved visual representation of the PDF, while the fitted result obtained is very similar to the original one.

While the histogram shape can be well defined given sufficient prior knowledge such as a control sample of data, or a sample of Monte Carlo simulated data, there are some

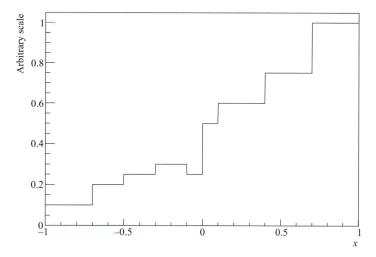

Figure B.8 An example of a histogram PDF.

instances when one does not have the luxury of an independent control sample to fix the PDF. One possible variant on the histogram PDF that can be used for such situation is the so-called parametric step function (PSF) (Aubert *et al.*, 2003). This is a PDF where data are binned in one or more dimensions, and the parameters of the PDF are the fractions of the total number of events in each bin. As binning can be non-uniform by definition, the fitted fractions are, in general, not the heights of each bin when plotted, but also depend on the bin width.

B.2.2 Kernal estimation PDF

A **kernal estimation** (KE) algorithm (for example, see Cranmer, 2001; Hastie, Tibshirani and Friedman, 2009) can be used to obtain a smoothly varying non-parametric PDF representation of a sample of data or Monte Carlo simulated data. The PDF is constructed from individual events used to represent a particular type of data, for example Monte Carlo simulated reference samples, or control samples of data. The individual events are referred to as kernals. Given a reference sample of data, one can define the properties of a kernal to be used in order to construct the PDF. A narrow step function would correspond to a shape similar to a histogram; however, a smoothly varying shape, such as a Gaussian could be used to provide a smooth non-parametric shape. For such a PDF the mean of the Gaussian is given by the point in space of ith event $e_i(\underline{x})$, and the width of the Gaussian kernal is taken from some representation of the data, for example the RMS of the data, some multiple of the RMS, or the RMS of a local set of the data. Given that each event in the reference sample contributes to the total PDF, one has to numerically integrate a KE PDF in order to normalise the total probability to unity.

A KE PDF for some data set Ω with N elements, with Gaussian kernals is given by

$$\mathcal{P}(x) = A \sum_{i=1}^{N} \frac{1}{\lambda\sqrt{2\pi}} e^{-(x-x_i)^2/2\lambda^2} , \tag{B.26}$$

where A is a normalisation coefficient, and λ is the kernal width, which is related to the σ of the data. In general $\lambda = a\sigma$, where a is a scale factor. This is set to one in order to use a kernal width corresponding to the RMS of Ω. The kernals can be made narrower ($a < 1$) or wider ($a > 1$) as desired in order to compute a PDF that is a better representation of the data. The use of different kernal elements is discussed in Hastie, Tibshirani and Friedman (2009).

The use case of the KE PDF is similar to that of the histogram PDF described above. The main difference between using a histogram PDF and a KE PDF is that the latter requires a lot more computation than the former. The reason for this is that a Gaussian kernal is computed for each event in the data set used to construct a KE PDF, and the PDF is evaluated at each point by computing a sum over all events. In practice it is usually beneficial to compute a KE PDF once and then store the output shape as a histogram, or in a look-up table, for use in all later fitting and validations. In essence the KE PDF is a smoothed representation of reference data. In contrast to a histogram PDF, where data are binned, the KE PDF constructs its representation of the data on an event-by-event basis.

KE PDFs are ideal for describing smoothly varying distributions that fall to zero at the extremities of the distribution. If this is not the case for a given situation, then one will encounter boundary effects that are associated with the fact that the value of the PDF at a given point in space x has a contribution from all events. Near boundaries, a PDF will obtain contributions only from events that are within the physical domain of the reference kernals being used to construct it, and hence the PDF will underestimate the distribution of events used to generate it. There are two ways to mitigate boundary effects in KE PDFs. Firstly the contributions from kernals near the boundary can be reflected at the boundary in order to compensate for missing contributions. This reflection or mirroring approach will tend to over-estimate the PDF at boundaries. A second possible way to mitigate boundary effects with a KE PDF is to provide events that go beyond the discriminating variable range required. This way the boundary effect is moved to the edges of the range of data being used, which is no longer in the valid range of the discriminating variable that one is trying to describe. Unfortunately the second approach is not always usable.

In general, KE PDFs are numerically resource hungry, and a minimum amount of kernal data is required in order to obtain a reasonable representation of the distribution. How much kernal data is required depends on the distribution being approximated, and the dimensionality of the problem. Thus far it has implicitly been assumed that the KE PDF is a one-dimensional PDF; however, one can extend the algorithm to an arbitrary number of dimensions. The penalties for doing this are that the computing power and the number of kernal data N required to compute the PDF increases considerably for each dimension added. This PDF suffers from the curse of dimensionality described by Bellman (1961).

B.2.3 Uniform distribution

This is a special case of the step function PDF described above and shown in Figure B.7. Consider the situation where the step function limits x_a and x_b coincide with the domain boundaries A and B in the problem. As a result of allowing the step function boundaries to match the limits of the interesting problem space, there are no intrinsic parameters required to determine the form of this PDF. Thus we can write

$$\mathcal{P}(x) = \frac{1}{B - A}, \tag{B.27}$$

where it can be shown that the normalisation condition for the PDF is satisfied as

$$\int\limits_{A}^{B} \mathcal{P}(x)dx = \int\limits_{A}^{B} \frac{1}{B-A}dx = 1. \tag{B.28}$$

If either A or B tends to infinity the normalisation of this PDF tends to zero. Hence, a uniform distribution is only well defined for some finite interval $x \in [A, B]$.

Appendix C

Numerical integration methods

This appendix provides a brief introduction to numerical integration techniques. A general approach that can be used for complicated distributions is to adopt a Monte Carlo method to compute a numerical integral. This follows from the discussion of how to compute an upper limit from an arbitrary PDF in Section 7.7. One draw back of this approach is that in order to obtain a precise estimate of an integral, one needs to generate a large number of simulated data. Thus if a PDF is easily integrable, it is highly inefficient to try and use Monte Carlo techniques to compute the integral. A number of robust numerical recipes are available to facilitate computation of integrals of single-valued functions. Two such algorithms are discussed in this appendix. With modern computers there can be less pressure on an individual to use efficient algorithms to compute numerical integrations; however, if an algorithm will be used on a regular basis, it would probably be beneficial to investigate the potential of more sophisticated algorithms than those presented here. Such techniques can be found in texts such as the Numerical Recipes series (Press *et al.*, 2002).

C.1 Rectangle rule

For some function $y = f(x)$, one can numerically determine an approximation for the integral

$$I = \int_{x_a}^{x_b} f(x)dx, \tag{C.1}$$

between the limits x_a and x_b. In the limit where dx becomes large, this integral approaches the approximation

$$I \simeq \sum_{i=1}^{n} f(x_i + \Delta x/2)\Delta x, \tag{C.2}$$

which is known as the ***rectangle rule***. Here Δx is the finite interval corresponding to the infinitesimal element dx and the sum is over n bins in x. Thus if we consider this sum graphically we represent the curve $f(x)$ by a set of discrete rectangles, each of some width Δx along the x direction, centred about the midpoint of a bin, $x + \Delta x/2$. This rectangle rule is also known as the midpoint rule, and it is the simplest approximation. If one requires a high level of accuracy in a numerical integral using this rule, then Δx must be small. As such the bin width Δx can be considered as a tuneable parameter.

By tuning or changing the value of Δx, one can estimate the numerical uncertainty on the use of this algorithm. With this information in hand for a given problem, the user can decide if a sufficient level of numerical precision has been achieved or not. One way of determining if a sufficient level of precision has been reached is to use the result of the numerical integral I_0 with the bin width Δx_0 to whatever end is desired (e.g. computation of a confidence level), and compare this with the corresponding result I_1 when the bin width is $\Delta x_1 = \Delta x_0/2$. If the result obtained in both cases is the same within the desired precision, then the numerical integration performed is sufficient. If not then the value of Δx is too large, and should be reduced further until the desired stability is encountered.

C.2 Trapezium rule

A more sophisticated numerical integration technique can be developed as a result of the lessons learned with the rectangle rule. Each contribution to the rectangle rule is an approximation of the curve that has no sensitivity to the change in gradient of the curve across a given bin i. One modification to this procedure would be to compute the numerical integral from a sum of trapeziums instead of rectangles. The error in this approximation should, in general (but not always), be smaller for each bin than for the rectangle rule, hence one would expect an algorithm based on a sum of trapeziums to compute a more accurate estimate of an integral than the rectangle rule for a given bin width.

The area of a trapezium A_i is given by

$$A_i = \frac{f(x_i) + f(x_{i+1})}{2} \Delta x. \tag{C.3}$$

If we consider n bins, then the corresponding area computed by the **trapezium rule** for integration is

$$A = \sum_{i=0}^{n} \left[\frac{f(x_i) + f(x_{i+1})}{2} \right] \Delta x. \tag{C.4}$$

The error associated with a particular integral can again be estimated by tuning the bin width Δx and observing how the computed integral impacts upon the end result.

C.3 Reflection on the use of numerical integration

Table C.1 summarises the results obtained on numerically integrating a Gaussian distribution between $\mu - 1\sigma$ and $\mu + 1\sigma$ for different numbers of bins, hence a number of different values of Δx. If we are content with obtaining a precision of three significant figures for our result, then the trapezium rule reaches this with 30 bins. In contrast, the rectangle rule achieves the same precision with only 10 bins and the integral is stable up to six significant figures when 900 000 bins are used with the rectangle rule. In this instance the more sophisticated algorithm provides a worse performance than the rectangle rule. If one considers the error computed on the integral, normalising to the result obtained with 10^6 bins, then it can be seen that the rectangle rule overestimates the integral and the trapezium rule underestimates the integral.

A second example is illustrated in Table C.2, where the results of performing a numerical integration of the cubic equation

$$y = x^3 + 2x^2 + 1, \tag{C.5}$$

are shown for $0 \le x \le 1$. Analytically this integral is $23/12$. For this particular function the rectangle rule is not as effective as the trapezium rule in computing an accurate estimate

Table C.1 *Comparison of numerical integrals of a Gaussian obtained using the rectangle and trapezium rules for different numbers of bins. These integrals are over the range* $\mu - 1\sigma$ *to* $\mu + 1\sigma$. *The third and fifth columns report the estimated error* (ϵ) *on the numerical integral for the two methods.*

Number of bins	Rectangle rule	ϵ	Trapezium rule	ϵ
5	0.685 946	3.3×10^{-3}	0.676 202	-6.5×10^{-3}
10	0.683 498	8.1×10^{-4}	0.681 074	-1.6×10^{-3}
50	0.682 722	3.3×10^{-5}	0.682 625	-6.4×10^{-5}
100	0.682 698	8.6×10^{-6}	0.682 673	-1.6×10^{-5}
10^6	0.682 689	–	0.682 689	–

Table C.2 *Comparison of numerical integrals of the cubic equation, Eq. (C.5), obtained using the rectangle and trapezium rules for different numbers of bins. These integrals are over the range* $0 \leq x \leq 1$. *The third and fifth columns report the estimated error* (ϵ) *on the numerical integral for the two methods.*

Number of bins	Rectangle rule	ϵ	Trapezium rule	ϵ
5	1.640 00	-0.277	1.940 00	0.023
10	1.772 50	-0.144	1.922 50	0.006
50	1.886 90	-0.030	1.916 90	2.3×10^{-4}
100	1.901 72	-0.015	1.916 72	5.8×10^{-5}
10^6	1.916 67	-1.5×10^{-7}	1.916 67	-1.5×10^{-13}

of the integral. The integral is stable up to six significant figures when one million steps are used. These two examples illustrate that the user should investigate the suitability and uncertainty obtained with different numerical techniques for each particular problem faced, as the effectiveness of a given numerical integration algorithm can change from case to case. It should be noted that if it is possible to perform the integral analytically then one should endeavour to do so.

One last thing to consider about numerical integration techniques is that the error on a computation is related to how quickly the gradient is changing over the width of a single bin. If you have to numerically integrate a function that has well-defined structure where parts of the function are smoothly varying and others change rapidly, then there may be advantages in trying to use an adaptive bin width. For example, the error penalty for using a large bin width when the function is varying smoothly may be relatively small, and by using a large bin width for such areas of the function you may be able to spend more CPU time computing the integral with a small bin width where that is more appropriate.

Appendix D

Solutions

2. Sets

2.1 (i) $A \cup B = \{1, 2, 3, 4, 5, 9\}$; (ii) $A \cap B = \{1, 3\}$; (iii) $A \setminus B = \{2, 4\}$; (iv) $B \setminus A = \{5, 9\}$; (v) $\bar{A} = \{5, 9\}$, and $\bar{B} = \{2, 4\}$

2.2 (i) $A \cup B = \{0, 1, 2, 3, 5, 6, 9\}$; (ii) $A \cap B = \{5\}$; (iii) $A \setminus B = \{0, 2, 6\}$; (iv) $B \setminus A = \{1, 3, 9\}$

2.3 (i) For example see the top part of Figure 2.3; (ii) see Figure 2.2, where A and B are interchanged.

2.4 $A \cup B = \mathbb{R}$

2.5 $A \cup B = \mathbb{R}$

2.6 $A \cap B = 1, 2, 3, 4, 5$

2.7 $A \cap B = \emptyset$

2.8 $A \cup B \cap C = \mathbb{Z}^{+}$

2.9 $A \cup B \cap \bar{C} = \{0, \mathbb{Z}^{-}\}$

2.10 $\Omega_{\text{Decimal}} = \{0, 1, 2, 3, 4, 5, 6, 7, 8, 9\}$

2.11 {red, green, blue}

2.12 $\mathbb{C}$

2.13 $\mathbb{Z}$

2.14 $\{x | 5 < x < 10\}$

2.15 $\{x | x \in \mathbb{R}\}$

3. Probability

3.1 $4/52 = 0.077$ to 3 s.f.

3.2 0.45%

3.3 $20/52 = 0.385$ to 3 s.f.

3.4 The probability of the getting an ace followed by a 10 point card, where the other player also gets a 10 point card is 0.72%. The probability of getting an ace followed by a 10 point card, where the other player gets a card with a different value is 1.69%. The total probability of being dealt an ace followed by a 10 point card is the sum: 2.41%.

3.5 The probability of rain tomorrow is 55.6%.

3.6 73.4%

3.7 The probability that it will rain tomorrow, using the prior that it rains 100 days of the year is 84.1%. The probability assuming that it is equally likely to rain or not is 93.3%. In both cases one concludes it is more probable that it will rain tomorrow.

3.8 The probability that it will rain tomorrow, using the prior that it rains 30 days of the year is 61.7%. The probability assuming that it is equally likely to rain or not is 94.7%. In both cases one concludes it is more probable that it will rain tomorrow, but in the first case the probability that it will rain is only slightly better than a 50/50 chance. In the second scenario the outcome of rain is predicted to be almost certain.

3.9 50%

3.10 To maximise the probability of winning the car, change your selection as the probability to win if you change is 2/3. The probability to win if you do not change is 1/3.

3.11 $N = 1/72$, and the likelihood is $\mathcal{L} = (3 - x)^2/9$ as the most probable value is at $x = 0$.

3.12 $N = 3/20$, and the likelihood is $\mathcal{L} = (1 + x + x^2)/7$ as the most probable value is at $x = 2$.

3.13 $N = 0.454\,597$, so $P(t) = 0.454\,597 e^{-t/2.2}$ and $\mathcal{L}(t) = e^{-t/2.2}$, where t is in μs, as the most probable value is at $t = 0$.

3.14 $\mathcal{L}(x = 1)/\mathcal{L}(x = 2) = e = 2.7182$ to 5 significant figures.

3.15 $\mathcal{L}(x = -1)/\mathcal{L}(x = 1) = 1$.

4. Visualising and quantifying the properties of data

4.1 $\mu_\Omega = 3.458$, $\sigma_\Omega = 2.635$, $\sigma_\Omega^2 = 6.941$, $\gamma_\Omega = 0.607$, $\mu_\kappa = 1.875$, $\sigma_\kappa = 0.872$, $\sigma_\kappa^2 = 0.760$, and $\gamma_\kappa = 0.290$. Note that as the number of events is small, one should use the definition of variance that includes the Bessel correction factor.

4.2

$$V = \begin{pmatrix} 6.941 & 2.236 \\ 2.236 & 0.760 \end{pmatrix} \tag{D.1}$$

4.3

$$\rho = \begin{pmatrix} 1 & 0.973 \\ 0.973 & 1 \end{pmatrix} \tag{D.2}$$

4.4 The eigenvalues are $\lambda_+ = 7.665$, and $\lambda_- = 0.036$, and the normalised eigenvectors are

$$\underline{x}_+ = \begin{pmatrix} 0.951 \\ 0.308 \end{pmatrix}, \text{ and } \underline{x}_- = \begin{pmatrix} 0.308 \\ -0.951 \end{pmatrix}. \tag{D.3}$$

Thus, the diagonalised form of the error matrix is

$$U = \begin{pmatrix} 7.665 & 0 \\ 0 & 0.036 \end{pmatrix}. \tag{D.4}$$

4.5 $\bar{x} = 3.5$, $\sigma_x^2 = 3.0$, $\sigma_x = 1.73$, and $\gamma = 0.0$, where the computation requires the use of variance with a Bessel correction factor to obtain an unbiased variance as the number of data is small.

4.6 $\bar{x} = 1.74$, $\sigma_x^2 = 0.743$, $\sigma_x = 0.862$, and $\gamma = 0.027$, where the computation requires the use of variance with a Bessel correction factor to obtain an unbiased variance as the number of data is small.

4.7 $\bar{x} = 0.659$, $\bar{y} = 1.300$, $\bar{z} = 0.118$, $\sigma(x) = 0.364$, $\sigma(y) = 0.184$, $\sigma(z) = 0.357$, and

$$\rho = \begin{pmatrix} 1 & 0.910 & 0.146 \\ 0.910 & 1 & 0.258 \\ 0.146 & 0.258 & 1 \end{pmatrix}. \tag{D.5}$$

4.8 $\bar{x} = 0.618$, $\bar{y} = 1.300$, $\bar{z} = 0.164$, $\sigma(x) = 0.360$, $\sigma(y) = 0.200$, $\sigma(z) = 0.238$, and

$$\rho = \begin{pmatrix} 1 & 0.833 & 0.733 \\ 0.833 & 1 & 0.757 \\ 0.733 & 0.757 & 1 \end{pmatrix}. \tag{D.6}$$

4.9 $\rho = 0.586$

4.10 $\rho = -0.014$

4.11 $\sigma_u^2 = (\sigma_x^2 \cos^2 \theta - \sigma_y^2 \sin^2 \theta)/(\cos^2 \theta - \sin^2 \theta)$, $\qquad \sigma_v^2 = (\sigma_y^2 \cos^2 \theta - \sigma_x^2 \sin^2 \theta)/$
$(\cos^2 \theta - \sin^2 \theta)$, and $\theta = \arctan(2\sigma_{xy}/(\sigma_x^2 - \sigma_y^2))/2$

4.12

$$U = \begin{pmatrix} 1.2 & 0.0 \\ 0.0 & 0.8 \end{pmatrix} \tag{D.7}$$

4.13

$$U = \begin{pmatrix} 2.21 & 0.00 \\ 0.00 & 0.79 \end{pmatrix} \tag{D.8}$$

4.14 The rotation angle is $\theta = 16.8°$, and the rotation matrix is (to 3 s.f.)

$$\begin{pmatrix} 0.957 & 0.290 \\ -0.290 & 0.957 \end{pmatrix}. \tag{D.9}$$

4.15 The rotation angle is $\theta = 24.4°$, and the rotation matrix is (to 3 s.f.)

$$\begin{pmatrix} 0.911 & 0.413 \\ -0.413 & 0.911 \end{pmatrix}. \tag{D.10}$$

5. Useful distributions

5.1 $P(5H) = P(0H) = 0.031\,25$, so $P(5H \vee 0H) = 0.062\,50$

5.2 $P(5H) = 0.010\,24$, $P(0H) = 0.077\,76$, so $P(5H \vee 0H) = 0.088$

5.3 0.2304

5.4 $P(0; 3) = 0.0498$

5.5 $P(\geq 3; 3) = 0.5768$

5.6 0.4688

5.7 $P(|x| \leq 1\sigma) = 0.6827$, $P(|x| \leq 2\sigma) = 0.9545$, and $P(|x| \leq 3\sigma) = 0.9973$

5.8 $z = 1.28$

5.9 0.6065

5.10 $P(\chi^2, \nu) = 0.2873$, this result is reasonable.

5.11 $P(\chi^2, \nu) = 0.9998$, this result is reasonable.

5.12 $P(\chi^2, \nu) = 0.0183$, this result is not reasonable.

5.13 mean = variance = $\lambda_1 + \lambda_2$. From this it follows that the sum of two Poisson distributions is a Poisson distribution.

5.14 $\frac{a}{2} - \frac{2}{e} + 1 \simeq \frac{a}{2} + 0.264$

5.15 $1 - \frac{2a}{3} - \frac{a^2}{4} - \frac{4}{e^2} - \frac{1}{e} + \frac{2a}{e}$

6. Uncertainty and errors

6.1 99.73%

6.2 0.663 ± 0.022

6.3 $\sigma_x = 0.1$, and $\sigma_y = 0.2$

6.4 $\sigma_A = 2\sqrt{(\epsilon[1-\epsilon])/N}$, where $\epsilon = N_1/N$

6.5 9.7 ± 2.4 m/s^2

6.6 Neglecting the uncertainty on L (which in this case is known to better than 1%), one expects a factor of 10 improvement in precision on g.

6.7 10.2 ± 0.6 m/s^2

6.8 $x = 1.5 \pm 0.2$

6.9 $x = 1.5 \pm 0.4$

6.10 $\sigma(Q) = \sqrt{(1-2\omega)^2 \sigma_\epsilon^2 + 4\epsilon^2 \sigma_\omega^2}$

6.11 Possible systematic uncertainties might include (i) student posture while being measured (ii) systematic offset from footwear of student (iii) accuracy of measuring device (iv) use of measuring device.

6.12 $A = 1.82 \pm 0.17$, $B = -0.74 \pm 0.26$ and cov$_{AB} = 0.013$.

6.13 $A = 1.59 \pm 0.35$, $B = 0.11 \pm 0.09$ and cov$_{AB} = 0.003$.

6.14 $A = 2.5 \pm 0.7$, $B = 0.2 \pm 0.9$ and $C = 0.5 \pm 0.7$.

6.15 $\sigma_{m_0}^2 = (E^4/c^4 + p^4)/(E^2 - p^2 c^2) \times 10^{-4}$

7. Confidence intervals

7.1 50% coverage: x within $\pm 0.67\sigma$ of the mean. 75% coverage: x within $\pm 1.15\sigma$ of the mean. 90% coverage: x within $\pm 1.64\sigma$ of the mean. 99% coverage: x within $\pm 2.58\sigma$ of the mean.

7.2 < 2.92 at 90% CL

7.3 $x < 1.19$ at 90% CL

7.4 Less than three events corresponds to a coverage of 98.1%, which is more than desired. If one linearly interpolates between $r = 2$ and $r = 3$, one obtains an estimate of the limit as 2.49.

7.5 $x < 1.9$ at 90% CL upper limit (statistical only); $x < 2.0$ 90% CL upper limit (statistical+systematic); $\Delta x_{UL} = 0.1$.

7.6 Mode is $\lambda = 3$, and from Figure 7.3, estimate $\lambda \in [1.6, 5.1]$ at 68.3 CL.

7.7 95% CL upper limit is approx. 7.8 events. This contains the two-sided interval, so one should either quote just the limit, or both interval and limit, with respective confidence levels.

7.8 $r \leq 7$ satisfies the coverage requirement, but at the 94.53% CL. Interpolating (linearly) between 6 and 7, one can estimate a 90% CL that is approximately 6.6.

7.9 $p < 0.189$ at 90% CL

7.10 $\epsilon > 0.977$ at 90% CL

7.11 $x < 2.303$ at 90% CL

7.12 $\lambda < 2.996$ at 95% CL

7.13 $S < -0.615$ at 90% CL

7.14 $\mu < 2.126$ at 90% CL

7.15 $N \in [1, 7]$ at 90% CL (actually this is over covered, the true coverage is 93.06%).

8. Hypothesis testing

8.1 The product of probability ratios of hypothesis A to hypothesis B is $R = 0.019$, hence hypothesis B is preferred by the data.

8.2 *P(infected | positive test result)* = 0.164. This is not a good test.

8.3 *P(infected | positive test result)* = 0.909. This is not a good test.

8.4 *P(infected | positive test result)* = 0.999 991. This is a good test.

8.5 The mean and standard deviation of the measurements are 1.85 m and 0.02 m, respectively. Given this the results appear to be in agreement with each other — but the quoted uncertainty of 0.01 is a factor of two smaller than the standard deviation. This could indicate the need to account for systematic uncertainties that have been neglected.

8.6 $\lambda = 3$ is preferred.

8.7 $\lambda = 4$ is preferred.

8.8 The Gaussian hypothesis is preferred.

8.9 The *p*-value is 0.0037, and is just compatible with expectations within 99.73% (Gaussian 3σ coverage level).

8.10 The *p*-value is 0.1847, and is compatible with expectations.

8.11 The *p*-value is 0.0002, and is not compatible with expectations at 99.73% (Gaussian 3σ coverage level).

8.12 $p = 10^{-6}$

8.13 $\Delta/\sigma_\Delta = 0.38$, the results agree.

8.14 $\Delta/\sigma_\Delta = 0.82$, the results agree.

8.15 Yes, both results can be correct. The first result would have drawn a conclusion corresponding to a false negative by placing an upper limit at approximately 1.28σ above the mean value for that result. The second measurement has a sufficiently large uncertainty that it would be compatible with the central value obtained by the first measurement. In fact, based solely on the second result, one would assume that 6.68% of repeat experiments with the same sensitivity might obtain a result ≤ 0 (assuming the method allowed one to obtain an un-physical result).

9. Fitting

9.1 It may be useful to fit data (i) in order to extract the values of one or more parameters (with uncertainties) (ii) obtain a (multi-dimensional) confidence interval (iii) average results.

9.2 The number of degrees of freedom ν is the total number of data minus the total number of constraints on the data. If there are no external constraints imposed, then the total number of data constitutes the only constraint on the problem and $\nu = N - 1$.

9.3 From that scan one finds that the mean value is 1.5, and the error is approximately 0.2 (given by a change of one in the χ^2 from the minimum).

9.4

$$a = \frac{\overline{yx^2} - \overline{y}\,\overline{x^2}}{\overline{x^4} - (\overline{x^2})^2}, \qquad b = \overline{y} - a\overline{x^2}.$$

9.5 $a = 1.975 \pm 0.029$, and $b = -0.047 \pm 0.091$.

9.6 The minimum is approximately $S = 0.66^{+0.025}_{-0.020}$.

9.7 The minimum is approximately $x = 1.25$, and the uncertainty on that value is approximately $+0.10 - 0.08$. The minimum $\chi^2 = 0.8125$ for $\nu = 1$, hence $P(\chi^2, \nu) = 0.37$. The value obtained is reasonable.

9.8 1.30 ± 0.67

9.9 -2.0 ± 1.0

9.10 1.0 ± 0.7 (1 d.p.)

9.11 $k = 1.96 \text{ N/cm}$
9.12 $R = \overline{IV}/\overline{I}^2$

10. Multivariate analysis

10.1 $\alpha \propto (-1.0, 1.0)$
10.2 $\alpha \propto (0.5, -0.5)$
10.3 $\alpha \propto (-1.6, -2.0, -0.2)$
10.4 $\alpha \propto (1.36, -1.09, 0.33)$
10.5 e: $E/p = 1.0, 0.9, 0.8$; π: $E/p = 0.7, 0.6, 0.5, 0.4, 0.3$; μ: $E/p = 0.2, 0.1, 0.0$.
10.6 Signal: $\omega_i = 0, 1$; background: $\omega_i = -3, -2, 4, 5$.
10.7 A: events 1 and 5; B: events 2, 3, and 4; A or B: event 6.
10.8 A: events 1, 4 and 5; B: events 2, 3, and 6.
10.9 The optimal cut value is 1.5.
10.10 The yields appear as a scale factor multiplying a ratio of integrals, and do not affect the optimal cut value obtained.
10.11 $S = 0.8$
10.12 $S = 0.5$
10.13 The optimal cut value is 0.75.
10.14 The optimal cut value is 8/9.
10.15 Simultaneously optimise upper and lower bounds on x in order to determine a cut interval $[X_L, X_U]$.

Appendix E

Reference tables

This appendix provides a number of quick-reference tables that can be used to compute probabilities based on the distributions introduced in Chapter 5. A more extensive set of information can be found in Lindley and Scott (1995).

E.1 Binomial probability

Section 5.2 introduced the binomial probability distribution, given by

$$P(r; p, n) = p^r (1 - p)^{n-r} \frac{n!}{r!(n - r)!},\tag{E.1}$$

for a given number of trials n with r successful outcomes, where each successful outcome has a probability given by p. Tables E.1 through E.5 summarise the cumulative probabilities for binomial distributions with $n = 2, 3, 4, 5,$ and 10, and p values between 0.1 and 0.9.

Table E.1 *Cumulative probability table for a binomial distribution with values of p specified at the top of each column, summing up from $r = 0$ to the r value specified in the first column. This table is for $n = 2$.*

	0.1	0.2	0.3	0.4	0.5	0.6	0.7	0.8	0.9
0	0.8100	0.6400	0.4900	0.3600	0.2500	0.1600	0.0900	0.0400	0.0100
1	0.9900	0.9600	0.9100	0.8400	0.7500	0.6400	0.5100	0.3600	0.1900
2	1.0000	1.0000	1.0000	1.0000	1.0000	1.0000	1.0000	1.0000	1.0000

Table E.2 *Cumulative probability table for a binomial distribution with values of p specified at the top of each column, summing up from $r = 0$ to the r value specified in the first column. This table is for $n = 3$.*

	0.1	0.2	0.3	0.4	0.5	0.6	0.7	0.8	0.9
0	0.7290	0.5120	0.3430	0.2160	0.1250	0.0640	0.0270	0.0080	0.0010
1	0.9720	0.8960	0.7840	0.6480	0.5000	0.3520	0.2160	0.1040	0.0280
2	0.9990	0.9920	0.9730	0.9360	0.8750	0.7840	0.6570	0.4880	0.2710
3	1.0000	1.0000	1.0000	1.0000	1.0000	1.0000	1.0000	1.0000	1.0000

Table E.3 *Cumulative probability table for a binomial distribution with values of p specified at the top of each column, summing up from r = 0 to the r value specified in the first column. This table is for n = 4.*

	0.1	0.2	0.3	0.4	0.5	0.6	0.7	0.8	0.9
0	0.6561	0.4096	0.2401	0.1296	0.0625	0.0256	0.0081	0.0016	0.0001
1	0.9477	0.8192	0.6517	0.4752	0.3125	0.1792	0.0837	0.0272	0.0037
2	0.9963	0.9728	0.9163	0.8208	0.6875	0.5248	0.3483	0.1808	0.0523
3	0.9999	0.9984	0.9919	0.9744	0.9375	0.8704	0.7599	0.5904	0.3439
4	1.0000	1.0000	1.0000	1.0000	1.0000	1.0000	1.0000	1.0000	1.0000

Table E.4 *Cumulative probability table for a binomial distribution with values of p specified at the top of each column, summing up from r = 0 to the r value specified in the first column. This table is for n = 5.*

	0.1	0.2	0.3	0.4	0.5	0.6	0.7	0.8	0.9
0	0.5905	0.3277	0.1681	0.0778	0.0312	0.0102	0.0024	0.0003	0.0000
1	0.9185	0.7373	0.5282	0.3370	0.1875	0.0870	0.0308	0.0067	0.0005
2	0.9914	0.9421	0.8369	0.6826	0.5000	0.3174	0.1631	0.0579	0.0086
3	0.9995	0.9933	0.9692	0.9130	0.8125	0.6630	0.4718	0.2627	0.0815
4	1.0000	0.9997	0.9976	0.9898	0.9688	0.9222	0.8319	0.6723	0.4095
5	1.0000	1.0000	1.0000	1.0000	1.0000	1.0000	1.0000	1.0000	1.0000

Table E.5 *Cumulative probability table for a binomial distribution with values of p specified at the top of each column, summing up from r = 0 to the r value specified in the first column. This table is for n = 10.*

	0.1	0.2	0.3	0.4	0.5	0.6	0.7	0.8	0.9
0	0.3487	0.1074	0.0282	0.0060	0.0010	0.0001	0.0000	0.0000	0.0000
1	0.7361	0.3758	0.1493	0.0464	0.0107	0.0017	0.0001	0.0000	0.0000
2	0.9298	0.6778	0.3828	0.1673	0.0547	0.0123	0.0016	0.0001	0.0000
3	0.9872	0.8791	0.6496	0.3823	0.1719	0.0548	0.0106	0.0009	0.0000
4	0.9984	0.9672	0.8497	0.6331	0.3770	0.1662	0.0473	0.0064	0.0001
5	0.9999	0.9936	0.9527	0.8338	0.6230	0.3669	0.1503	0.0328	0.0016
6	1.0000	0.9991	0.9894	0.9452	0.8281	0.6177	0.3504	0.1209	0.0128
7	1.0000	0.9999	0.9984	0.9877	0.9453	0.8327	0.6172	0.3222	0.0702
8	1.0000	1.0000	0.9999	0.9983	0.9893	0.9536	0.8507	0.6242	0.2639
9	1.0000	1.0000	1.0000	0.9999	0.9990	0.9940	0.9718	0.8926	0.6513
10	1.0000	1.0000	1.0000	1.0000	1.0000	1.0000	1.0000	1.0000	1.0000

Table E.6 *Probability table for a Poisson distribution $P(r, \lambda)$ with values of λ specified at the top of each column, for a given value of r specified in the first column.*

	1	2	3	4	5	6	7	8	9	10
0	0.3679	0.1353	0.0498	0.0183	0.0067	0.0025	0.0009	0.0003	0.0001	0.0000
1	0.3679	0.2707	0.1494	0.0733	0.0337	0.0149	0.0064	0.0027	0.0011	0.0005
2	0.1839	0.2707	0.2240	0.1465	0.0842	0.0446	0.0223	0.0107	0.0050	0.0023
3	0.0613	0.1804	0.2240	0.1954	0.1404	0.0892	0.0521	0.0286	0.0150	0.0076
4	0.0153	0.0902	0.1680	0.1954	0.1755	0.1339	0.0912	0.0573	0.0337	0.0189
5	0.0031	0.0361	0.1008	0.1563	0.1755	0.1606	0.1277	0.0916	0.0607	0.0378
6	0.0005	0.0120	0.0504	0.1042	0.1462	0.1606	0.1490	0.1221	0.0911	0.0631
7	0.0001	0.0034	0.0216	0.0595	0.1044	0.1377	0.1490	0.1396	0.1171	0.0901
8	0.0000	0.0009	0.0081	0.0298	0.0653	0.1033	0.1304	0.1396	0.1318	0.1126
9	0.0000	0.0002	0.0027	0.0132	0.0363	0.0688	0.1014	0.1241	0.1318	0.1251
10	0.0000	0.0000	0.0008	0.0053	0.0181	0.0413	0.0710	0.0993	0.1186	0.1251
11	0.0000	0.0000	0.0002	0.0019	0.0082	0.0225	0.0452	0.0722	0.0970	0.1137
12	0.0000	0.0000	0.0001	0.0006	0.0034	0.0113	0.0263	0.0481	0.0728	0.0948
13	0.0000	0.0000	0.0000	0.0002	0.0013	0.0052	0.0142	0.0296	0.0504	0.0729
14	0.0000	0.0000	0.0000	0.0001	0.0005	0.0022	0.0071	0.0169	0.0324	0.0521
15	0.0000	0.0000	0.0000	0.0000	0.0002	0.0009	0.0033	0.0090	0.0194	0.0347
16	0.0000	0.0000	0.0000	0.0000	0.0000	0.0003	0.0014	0.0045	0.0109	0.0217
17	0.0000	0.0000	0.0000	0.0000	0.0000	0.0001	0.0006	0.0021	0.0058	0.0128
18	0.0000	0.0000	0.0000	0.0000	0.0000	0.0000	0.0002	0.0009	0.0029	0.0071
19	0.0000	0.0000	0.0000	0.0000	0.0000	0.0000	0.0001	0.0004	0.0014	0.0037
20	0.0000	0.0000	0.0000	0.0000	0.0000	0.0000	0.0000	0.0002	0.0006	0.0019

E.2 Poisson probability

Section 5.3 introduced the Poisson probability distribution, given by

$$P(r, \lambda) = \frac{\lambda^r e^{-\lambda}}{r!}. \tag{E.2}$$

Poisson probabilities $P(r, \lambda)$ for integer values of λ from one to ten and values of r up to twenty are summarised in Table E.6. The first column indicates the value of r, and the top row indicates the value of λ considered.

The corresponding table for cumulative probabilities (summing up from $r = 0$ to the specified value) given by

$$P = \sum_{r=0}^{r=n} P(r, \lambda), \tag{E.3}$$

can be found in Table E.7.

Table E.7 *Cumulative probability table for a Poisson distribution with values of* λ *specified at the top of each column, summing up from* $r = 0$ *to the* r *value specified in the first column.*

	1	2	3	4	5	6	7	8	9	10
0	0.3679	0.1353	0.0498	0.0183	0.0067	0.0025	0.0009	0.0003	0.0001	0.0000
1	0.7358	0.4060	0.1991	0.0916	0.0404	0.0174	0.0073	0.0030	0.0012	0.0005
2	0.9197	0.6767	0.4232	0.2381	0.1247	0.0620	0.0296	0.0138	0.0062	0.0028
3	0.9810	0.8571	0.6472	0.4335	0.2650	0.1512	0.0818	0.0424	0.0212	0.0103
4	0.9963	0.9473	0.8153	0.6288	0.4405	0.2851	0.1730	0.0996	0.0550	0.0293
5	0.9994	0.9834	0.9161	0.7851	0.6160	0.4457	0.3007	0.1912	0.1157	0.0671
6	0.9999	0.9955	0.9665	0.8893	0.7622	0.6063	0.4497	0.3134	0.2068	0.1301
7	1.0000	0.9989	0.9881	0.9489	0.8666	0.7440	0.5987	0.4530	0.3239	0.2202
8	1.0000	0.9998	0.9962	0.9786	0.9319	0.8472	0.7291	0.5925	0.4557	0.3328
9	1.0000	1.0000	0.9989	0.9919	0.9682	0.9161	0.8305	0.7166	0.5874	0.4579
10	1.0000	1.0000	0.9997	0.9972	0.9863	0.9574	0.9015	0.8159	0.7060	0.5830
11	1.0000	1.0000	0.9999	0.9991	0.9945	0.9799	0.9467	0.8881	0.8030	0.6968
12	1.0000	1.0000	1.0000	0.9997	0.9980	0.9912	0.9730	0.9362	0.8758	0.7916
13	1.0000	1.0000	1.0000	0.9999	0.9993	0.9964	0.9872	0.9658	0.9261	0.8645
14	1.0000	1.0000	1.0000	1.0000	0.9998	0.9986	0.9943	0.9827	0.9585	0.9165
15	1.0000	1.0000	1.0000	1.0000	0.9999	0.9995	0.9976	0.9918	0.9780	0.9513
16	1.0000	1.0000	1.0000	1.0000	1.0000	0.9998	0.9990	0.9963	0.9889	0.9730
17	1.0000	1.0000	1.0000	1.0000	1.0000	0.9999	0.9996	0.9984	0.9947	0.9857
18	1.0000	1.0000	1.0000	1.0000	1.0000	1.0000	0.9999	0.9993	0.9976	0.9928
19	1.0000	1.0000	1.0000	1.0000	1.0000	1.0000	1.0000	0.9997	0.9989	0.9965
20	1.0000	1.0000	1.0000	1.0000	1.0000	1.0000	1.0000	0.9999	0.9996	0.9984

E.3 Gaussian probability

Section 5.4 introduced the Gaussian probability distribution, which is given by

$$P(x, \mu, \sigma) = \frac{1}{\sigma\sqrt{2\pi}} e^{-(x-\mu)^2/2\sigma^2}. \tag{E.4}$$

One- and two-sided confidence intervals for the Gaussian probability distribution can be found in Tables E.8 and E.9. The one-sided confidence interval is given by

$$P = \int_{-\infty}^{n\sigma} \frac{1}{\sigma\sqrt{2\pi}} e^{-(x-\mu)^2/2\sigma^2} dx, \tag{E.5}$$

$$= \int_{-\infty}^{n} \frac{1}{\sqrt{2\pi}} e^{-z^2/2} dz, \tag{E.6}$$

where the lower limit is $-\infty$ and the upper limit is the number of σ above the mean value. The table is given in terms of $z = (x - \mu)/\sigma$, where the first column indicates the number of σ up to one decimal place, and the first row indicates the number of σ at the level of the second decimal place.

For example, if one is interested in the probability corresponding to a one-sided Gaussian interval from $-\infty$ to $\mu + 1.55\sigma$, then this would be given by 0.9394, which corresponds to the element in the 0.05 column and row marked with 1.5 in Table E.8.

Table E.8 *One-sided probability table for a Gaussian distribution in terms of the number of standard deviations from the mean value for $z > 0$.*

	0.00	0.01	0.02	0.03	0.04	0.05	0.06	0.07	0.08	0.09
0.0	0.5000	0.5040	0.5080	0.5120	0.5160	0.5199	0.5239	0.5279	0.5319	0.5359
0.1	0.5398	0.5438	0.5478	0.5517	0.5557	0.5596	0.5636	0.5675	0.5714	0.5753
0.2	0.5793	0.5832	0.5871	0.5910	0.5948	0.5987	0.6026	0.6064	0.6103	0.6141
0.3	0.6179	0.6217	0.6255	0.6293	0.6331	0.6368	0.6406	0.6443	0.6480	0.6517
0.4	0.6554	0.6591	0.6628	0.6664	0.6700	0.6736	0.6772	0.6808	0.6844	0.6879
0.5	0.6915	0.6950	0.6985	0.7019	0.7054	0.7088	0.7123	0.7157	0.7190	0.7224
0.6	0.7257	0.7291	0.7324	0.7357	0.7389	0.7422	0.7454	0.7486	0.7517	0.7549
0.7	0.7580	0.7611	0.7642	0.7673	0.7704	0.7734	0.7764	0.7794	0.7823	0.7852
0.8	0.7881	0.7910	0.7939	0.7967	0.7995	0.8023	0.8051	0.8078	0.8106	0.8133
0.9	0.8159	0.8186	0.8212	0.8238	0.8264	0.8289	0.8315	0.8340	0.8365	0.8389
1.0	0.8413	0.8438	0.8461	0.8485	0.8508	0.8531	0.8554	0.8577	0.8599	0.8621
1.1	0.8643	0.8665	0.8686	0.8708	0.8729	0.8749	0.8770	0.8790	0.8810	0.8830
1.2	0.8849	0.8869	0.8888	0.8907	0.8925	0.8944	0.8962	0.8980	0.8997	0.9015
1.3	0.9032	0.9049	0.9066	0.9082	0.9099	0.9115	0.9131	0.9147	0.9162	0.9177
1.4	0.9192	0.9207	0.9222	0.9236	0.9251	0.9265	0.9279	0.9292	0.9306	0.9319
1.5	0.9332	0.9345	0.9357	0.9370	0.9382	0.9394	0.9406	0.9418	0.9429	0.9441
1.6	0.9452	0.9463	0.9474	0.9484	0.9495	0.9505	0.9515	0.9525	0.9535	0.9545
1.7	0.9554	0.9564	0.9573	0.9582	0.9591	0.9599	0.9608	0.9616	0.9625	0.9633
1.8	0.9641	0.9649	0.9656	0.9664	0.9671	0.9678	0.9686	0.9693	0.9699	0.9706
1.9	0.9713	0.9719	0.9726	0.9732	0.9738	0.9744	0.9750	0.9756	0.9761	0.9767
2.0	0.9773	0.9778	0.9783	0.9788	0.9793	0.9798	0.9803	0.9808	0.9812	0.9817
2.1	0.9821	0.9826	0.9830	0.9834	0.9838	0.9842	0.9846	0.9850	0.9854	0.9857
2.2	0.9861	0.9864	0.9868	0.9871	0.9875	0.9878	0.9881	0.9884	0.9887	0.9890
2.3	0.9893	0.9896	0.9898	0.9901	0.9904	0.9906	0.9909	0.9911	0.9913	0.9916
2.4	0.9918	0.9920	0.9922	0.9925	0.9927	0.9929	0.9931	0.9932	0.9934	0.9936
2.5	0.9938	0.9940	0.9941	0.9943	0.9945	0.9946	0.9948	0.9949	0.9951	0.9952
2.6	0.9953	0.9955	0.9956	0.9957	0.9959	0.9960	0.9961	0.9962	0.9963	0.9964
2.7	0.9965	0.9966	0.9967	0.9968	0.9969	0.9970	0.9971	0.9972	0.9973	0.9974
2.8	0.9974	0.9975	0.9976	0.9977	0.9977	0.9978	0.9979	0.9979	0.9980	0.9981
2.9	0.9981	0.9982	0.9983	0.9983	0.9984	0.9984	0.9985	0.9985	0.9986	0.9986
3.0	0.9987	0.9987	0.9987	0.9988	0.9988	0.9989	0.9989	0.9989	0.9990	0.9990
3.1	0.9990	0.9991	0.9991	0.9991	0.9992	0.9992	0.9992	0.9992	0.9993	0.9993
3.2	0.9993	0.9993	0.9994	0.9994	0.9994	0.9994	0.9994	0.9995	0.9995	0.9995
3.3	0.9995	0.9995	0.9996	0.9996	0.9996	0.9996	0.9996	0.9996	0.9996	0.9997
3.4	0.9997	0.9997	0.9997	0.9997	0.9997	0.9997	0.9997	0.9997	0.9997	0.9998
3.5	0.9998	0.9998	0.9998	0.9998	0.9998	0.9998	0.9998	0.9998	0.9998	0.9998
3.6	0.9998	0.9998	0.9999	0.9999	0.9999	0.9999	0.9999	0.9999	0.9999	0.9999
3.7	0.9999	0.9999	0.9999	0.9999	0.9999	0.9999	0.9999	0.9999	0.9999	0.9999
3.8	0.9999	0.9999	0.9999	0.9999	0.9999	0.9999	0.9999	0.9999	0.9999	1.0000
3.9	1.0000	1.0000	1.0000	1.0000	1.0000	1.0000	1.0000	1.0000	1.0000	1.0000

Table E.9 *Two-sided probability table for a Gaussian distribution in terms of the number of standard deviations from the mean value.*

	0.00	0.01	0.02	0.03	0.04	0.05	0.06	0.07	0.08	0.09
0.0	0.0000	0.0080	0.0160	0.0239	0.0319	0.0399	0.0478	0.0558	0.0638	0.0717
0.1	0.0797	0.0876	0.0955	0.1034	0.1113	0.1192	0.1271	0.1350	0.1428	0.1507
0.2	0.1585	0.1663	0.1741	0.1819	0.1897	0.1974	0.2051	0.2128	0.2205	0.2282
0.3	0.2358	0.2434	0.2510	0.2586	0.2661	0.2737	0.2812	0.2886	0.2961	0.3035
0.4	0.3108	0.3182	0.3255	0.3328	0.3401	0.3473	0.3545	0.3616	0.3688	0.3759
0.5	0.3829	0.3899	0.3969	0.4039	0.4108	0.4177	0.4245	0.4313	0.4381	0.4448
0.6	0.4515	0.4581	0.4647	0.4713	0.4778	0.4843	0.4907	0.4971	0.5035	0.5098
0.7	0.5161	0.5223	0.5285	0.5346	0.5407	0.5467	0.5527	0.5587	0.5646	0.5705
0.8	0.5763	0.5821	0.5878	0.5935	0.5991	0.6047	0.6102	0.6157	0.6211	0.6265
0.9	0.6319	0.6372	0.6424	0.6476	0.6528	0.6579	0.6629	0.6680	0.6729	0.6778
1.0	0.6827	0.6875	0.6923	0.6970	0.7017	0.7063	0.7109	0.7154	0.7199	0.7243
1.1	0.7287	0.7330	0.7373	0.7415	0.7457	0.7499	0.7540	0.7580	0.7620	0.7660
1.2	0.7699	0.7737	0.7775	0.7813	0.7850	0.7887	0.7923	0.7959	0.7995	0.8029
1.3	0.8064	0.8098	0.8132	0.8165	0.8198	0.8230	0.8262	0.8293	0.8324	0.8355
1.4	0.8385	0.8415	0.8444	0.8473	0.8501	0.8529	0.8557	0.8584	0.8611	0.8638
1.5	0.8664	0.8690	0.8715	0.8740	0.8764	0.8789	0.8812	0.8836	0.8859	0.8882
1.6	0.8904	0.8926	0.8948	0.8969	0.8990	0.9011	0.9031	0.9051	0.9070	0.9090
1.7	0.9109	0.9127	0.9146	0.9164	0.9181	0.9199	0.9216	0.9233	0.9249	0.9265
1.8	0.9281	0.9297	0.9312	0.9328	0.9342	0.9357	0.9371	0.9385	0.9399	0.9412
1.9	0.9426	0.9439	0.9451	0.9464	0.9476	0.9488	0.9500	0.9512	0.9523	0.9534
2.0	0.9545	0.9556	0.9566	0.9576	0.9586	0.9596	0.9606	0.9615	0.9625	0.9634
2.1	0.9643	0.9651	0.9660	0.9668	0.9676	0.9684	0.9692	0.9700	0.9707	0.9715
2.2	0.9722	0.9729	0.9736	0.9743	0.9749	0.9756	0.9762	0.9768	0.9774	0.9780
2.3	0.9786	0.9791	0.9797	0.9802	0.9807	0.9812	0.9817	0.9822	0.9827	0.9832
2.4	0.9836	0.9840	0.9845	0.9849	0.9853	0.9857	0.9861	0.9865	0.9869	0.9872
2.5	0.9876	0.9879	0.9883	0.9886	0.9889	0.9892	0.9895	0.9898	0.9901	0.9904
2.6	0.9907	0.9909	0.9912	0.9915	0.9917	0.9920	0.9922	0.9924	0.9926	0.9929
2.7	0.9931	0.9933	0.9935	0.9937	0.9939	0.9940	0.9942	0.9944	0.9946	0.9947
2.8	0.9949	0.9950	0.9952	0.9953	0.9955	0.9956	0.9958	0.9959	0.9960	0.9961
2.9	0.9963	0.9964	0.9965	0.9966	0.9967	0.9968	0.9969	0.9970	0.9971	0.9972
3.0	0.9973	0.9974	0.9975	0.9976	0.9976	0.9977	0.9978	0.9979	0.9979	0.9980
3.1	0.9981	0.9981	0.9982	0.9983	0.9983	0.9984	0.9984	0.9985	0.9985	0.9986
3.2	0.9986	0.9987	0.9987	0.9988	0.9988	0.9988	0.9989	0.9989	0.9990	0.9990
3.3	0.9990	0.9991	0.9991	0.9991	0.9992	0.9992	0.9992	0.9992	0.9993	0.9993
3.4	0.9993	0.9994	0.9994	0.9994	0.9994	0.9994	0.9995	0.9995	0.9995	0.9995
3.5	0.9995	0.9996	0.9996	0.9996	0.9996	0.9996	0.9996	0.9996	0.9997	0.9997
3.6	0.9997	0.9997	0.9997	0.9997	0.9997	0.9997	0.9997	0.9998	0.9998	0.9998
3.7	0.9998	0.9998	0.9998	0.9998	0.9998	0.9998	0.9998	0.9998	0.9998	0.9998
3.8	0.9999	0.9999	0.9999	0.9999	0.9999	0.9999	0.9999	0.9999	0.9999	0.9999
3.9	0.9999	0.9999	0.9999	0.9999	0.9999	0.9999	0.9999	0.9999	0.9999	0.9999

Table E.10 χ^2 *probability table for observed values of (rows)* χ^2 *and (columns) degrees of freedom* ν.

	1	2	3	4	5	6	7	8	9	10
0.1	0.7518	0.9512	0.9918	0.9988	0.9998	1.0000	1.0000	1.0000	1.0000	1.0000
0.2	0.6547	0.9048	0.9776	0.9953	0.9991	0.9998	1.0000	1.0000	1.0000	1.0000
0.3	0.5839	0.8607	0.9600	0.9898	0.9976	0.9995	0.9999	1.0000	1.0000	1.0000
0.4	0.5271	0.8187	0.9402	0.9825	0.9953	0.9989	0.9997	0.9999	1.0000	1.0000
0.5	0.4795	0.7788	0.9189	0.9735	0.9921	0.9978	0.9994	0.9999	1.0000	1.0000
0.6	0.4386	0.7408	0.8964	0.9631	0.9880	0.9964	0.9990	0.9997	0.9999	1.0000
0.7	0.4028	0.7047	0.8732	0.9513	0.9830	0.9945	0.9983	0.9995	0.9999	1.0000
0.8	0.3711	0.6703	0.8495	0.9384	0.9770	0.9921	0.9974	0.9992	0.9998	0.9999
0.9	0.3428	0.6376	0.8254	0.9246	0.9702	0.9891	0.9963	0.9988	0.9996	0.9999
1.0	0.3173	0.6065	0.8013	0.9098	0.9626	0.9856	0.9948	0.9982	0.9994	0.9998
1.1	0.2943	0.5769	0.7771	0.8943	0.9541	0.9815	0.9931	0.9975	0.9992	0.9997
1.2	0.2733	0.5488	0.7530	0.8781	0.9449	0.9769	0.9909	0.9966	0.9988	0.9996
1.3	0.2542	0.5220	0.7291	0.8614	0.9349	0.9717	0.9884	0.9956	0.9984	0.9994
1.4	0.2367	0.4966	0.7055	0.8442	0.9243	0.9659	0.9856	0.9942	0.9978	0.9992
1.5	0.2207	0.4724	0.6823	0.8266	0.9131	0.9595	0.9823	0.9927	0.9971	0.9989
1.6	0.2059	0.4493	0.6594	0.8088	0.9012	0.9526	0.9786	0.9909	0.9963	0.9986
1.7	0.1923	0.4274	0.6369	0.7907	0.8889	0.9451	0.9746	0.9889	0.9954	0.9982
1.8	0.1797	0.4066	0.6149	0.7725	0.8761	0.9371	0.9701	0.9865	0.9942	0.9977
1.9	0.1681	0.3867	0.5934	0.7541	0.8628	0.9287	0.9652	0.9839	0.9930	0.9971
2.0	0.1573	0.3679	0.5724	0.7358	0.8491	0.9197	0.9598	0.9810	0.9915	0.9963
2.1	0.1473	0.3499	0.5519	0.7174	0.8351	0.9103	0.9541	0.9778	0.9898	0.9955
2.2	0.1380	0.3329	0.5319	0.6990	0.8208	0.9004	0.9479	0.9743	0.9879	0.9946
2.3	0.1294	0.3166	0.5125	0.6808	0.8063	0.8901	0.9414	0.9704	0.9858	0.9935
2.4	0.1213	0.3012	0.4936	0.6626	0.7915	0.8795	0.9344	0.9662	0.9835	0.9923
2.5	0.1138	0.2865	0.4753	0.6446	0.7765	0.8685	0.9271	0.9617	0.9809	0.9909
2.6	0.1069	0.2725	0.4575	0.6268	0.7614	0.8571	0.9194	0.9569	0.9781	0.9893
2.7	0.1003	0.2592	0.4402	0.6092	0.7461	0.8454	0.9113	0.9518	0.9750	0.9876
2.8	0.0943	0.2466	0.4235	0.5918	0.7308	0.8335	0.9029	0.9463	0.9717	0.9857
2.9	0.0886	0.2346	0.4073	0.5747	0.7154	0.8213	0.8941	0.9405	0.9681	0.9837

The two-sided confidence interval is given by

$$P = \int_{-n\sigma}^{n\sigma} \frac{1}{\sigma\sqrt{2\pi}} e^{-(x-\mu)^2/2\sigma^2} dx, \tag{E.7}$$

$$= \int_{-n}^{n} \frac{1}{\sqrt{2\pi}} e^{-z^2/2} dz, \tag{E.8}$$

where n defines the number of σ over which the integral is performed. For example, if one is interested in the probability corresponding to the two sided Gaussian interval between $\mu - 1.64\sigma$ and $\mu + 1.64\sigma$, then this would correspond to 0.8990, which is the table element in the 0.04 column and the row marked with 1.6 in Table E.9.

Table E.11 χ^2 *probability table for observed values of (rows) χ^2 and (columns) degrees of freedom ν.*

	1	2	3	4	5	6	7	8	9	10
0.5	0.4795	0.7788	0.9189	0.9735	0.9921	0.9978	0.9994	0.9999	1.0000	1.0000
1.0	0.3173	0.6065	0.8013	0.9098	0.9626	0.9856	0.9948	0.9982	0.9994	0.9998
1.5	0.2207	0.4724	0.6823	0.8266	0.9131	0.9595	0.9823	0.9927	0.9971	0.9989
2.0	0.1573	0.3679	0.5724	0.7358	0.8491	0.9197	0.9598	0.9810	0.9915	0.9963
2.5	0.1138	0.2865	0.4753	0.6446	0.7765	0.8685	0.9271	0.9617	0.9809	0.9909
3.0	0.0833	0.2231	0.3916	0.5578	0.7000	0.8088	0.8850	0.9344	0.9643	0.9814
3.5	0.0614	0.1738	0.3208	0.4779	0.6234	0.7440	0.8352	0.8992	0.9411	0.9671
4.0	0.0455	0.1353	0.2615	0.4060	0.5494	0.6767	0.7798	0.8571	0.9114	0.9473
4.5	0.0339	0.1054	0.2123	0.3425	0.4799	0.6093	0.7207	0.8094	0.8755	0.9220
5.0	0.0253	0.0821	0.1718	0.2873	0.4159	0.5438	0.6600	0.7576	0.8343	0.8912
5.5	0.0190	0.0639	0.1386	0.2397	0.3579	0.4815	0.5992	0.7030	0.7887	0.8554
6.0	0.0143	0.0498	0.1116	0.1991	0.3062	0.4232	0.5397	0.6472	0.7399	0.8153
6.5	0.0108	0.0388	0.0897	0.1648	0.2606	0.3696	0.4827	0.5914	0.6890	0.7717
7.0	0.0082	0.0302	0.0719	0.1359	0.2206	0.3208	0.4289	0.5366	0.6371	0.7254
7.5	0.0062	0.0235	0.0576	0.1117	0.1860	0.2771	0.3787	0.4838	0.5852	0.6775
8.0	0.0047	0.0183	0.0460	0.0916	0.1562	0.2381	0.3326	0.4335	0.5341	0.6288
8.5	0.0036	0.0143	0.0367	0.0749	0.1307	0.2037	0.2906	0.3862	0.4846	0.5801
9.0	0.0027	0.0111	0.0293	0.0611	0.1091	0.1736	0.2527	0.3423	0.4373	0.5321
9.5	0.0021	0.0087	0.0233	0.0497	0.0907	0.1473	0.2187	0.3019	0.3925	0.4854
10.0	0.0016	0.0067	0.0186	0.0404	0.0752	0.1247	0.1886	0.2650	0.3505	0.4405
10.5	0.0012	0.0052	0.0148	0.0328	0.0622	0.1051	0.1620	0.2317	0.3115	0.3978
11.0	0.0009	0.0041	0.0117	0.0266	0.0514	0.0884	0.1386	0.2017	0.2757	0.3575
11.5	0.0007	0.0032	0.0093	0.0215	0.0423	0.0741	0.1182	0.1749	0.2430	0.3199
12.0	0.0005	0.0025	0.0074	0.0174	0.0348	0.0620	0.1006	0.1512	0.2133	0.2851
12.5	0.0004	0.0019	0.0059	0.0140	0.0285	0.0517	0.0853	0.1303	0.1866	0.2530
13.0	0.0003	0.0015	0.0046	0.0113	0.0234	0.0430	0.0721	0.1118	0.1626	0.2237
13.5	0.0002	0.0012	0.0037	0.0091	0.0191	0.0357	0.0608	0.0958	0.1413	0.1970
14.0	0.0002	0.0009	0.0029	0.0073	0.0156	0.0296	0.0512	0.0818	0.1223	0.1730
14.5	0.0001	0.0007	0.0023	0.0059	0.0127	0.0245	0.0430	0.0696	0.1056	0.1514
15.0	0.0001	0.0006	0.0018	0.0047	0.0104	0.0203	0.0360	0.0591	0.0909	0.1321
15.5	0.0001	0.0004	0.0014	0.0038	0.0084	0.0167	0.0301	0.0501	0.0781	0.1149
16.0	0.0001	0.0003	0.0011	0.0030	0.0068	0.0138	0.0251	0.0424	0.0669	0.0996
16.5	0.0000	0.0003	0.0009	0.0024	0.0056	0.0113	0.0209	0.0358	0.0571	0.0862
17.0	0.0000	0.0002	0.0007	0.0019	0.0045	0.0093	0.0174	0.0301	0.0487	0.0744
17.5	0.0000	0.0002	0.0006	0.0015	0.0036	0.0076	0.0144	0.0253	0.0414	0.0640
18.0	0.0000	0.0001	0.0004	0.0012	0.0029	0.0062	0.0120	0.0212	0.0352	0.0550
18.5	0.0000	0.0001	0.0003	0.0010	0.0024	0.0051	0.0099	0.0178	0.0298	0.0471
19.0	0.0000	0.0001	0.0003	0.0008	0.0019	0.0042	0.0082	0.0149	0.0252	0.0403
19.5	0.0000	0.0001	0.0002	0.0006	0.0016	0.0034	0.0068	0.0124	0.0213	0.0344
20.0	0.0000	0.0000	0.0002	0.0005	0.0012	0.0028	0.0056	0.0103	0.0179	0.0293

E.4 χ^2 **probability**

The probability for obtaining a given χ^2 for different numbers of degrees of freedom can be found in Tables E.10 and E.11. The χ^2 probability for a given number of degrees of freedom is

$$P(\chi^2, v) = \frac{2^{-v/2}}{\Gamma(v/2)} (\chi^2)^{(v/2-1)} e^{-\chi^2/2},$$

(E.9)

and is discussed in Section 5.5.

References

Abe, K. *et al.* 2011, *Phys. Rev. Lett.*, **107**, 041801.

Aubert, B. *et al.* 2003, *Phys. Rev. Lett.*, **91**, 241801.

Aubert, B. *et al.* 2007, *Phys. Rev. D*, **76**, 052007.

Aubert, B. *et al.* 2009, *Phys. Rev. D*, **79**, 072009.

Barlow, R. 1989, *Statistics*, Wiley.

Barlow, R. 1990, *Nucl. Instrum. Meth. A*, **297**, 496–506.

Baum, E. and Haussler, D. 1989, *Neural Comp.*, **1**, 151–160.

Bayes, T. 1763, Reprinted in 1958, *Biometrika*, **45** (3–4), 296–315.

Bellman, R. 1961, *Adaptive Control Processes: A Guided Tour*, Oxford University Press.

Beringer, J. *et al.* (Particle Data Group) 2012, *Phys. Rev. D*, **86**, 010001.

Cousins, R. D. and Highland, V. L. 1992, *Nucl. Intrum. Meth A*, **320**, 331–335.

Cowan, G. 1998, *Statistical Data Analysis*, Oxford University Press.

Cranmer, K. S. 2001, *Comp. Phys. Comm.*, **136**, 198.

Davidson, A. 2003, *Statistical Models*, Cambridge University Press.

Dunnington, F. 1933, *Phys. Rev.*, **43**, 404.

Edwards, A. 1992, *Likelihood*, John Hopkins University Press.

Fanchiotti, H., Garcia Canal, C. A. and Marucho, M. 2006, *Int. J. Mod. Phys.* C**17** 1461–1476.

Feldman, G. and Cousins, R. 1998, *Phys. Rev. D*, **57**, 3873–3889.

Feynman, R. P., Leighton, R. B. and Sands, M. L. 1989, *The Feynman Lectures on Physics Volume I*, Addison-Wesley.

Fisher, R. A. 1936, *Ann. Eug.*, **7**, 179–188.

Gaiser, J. E. 1982, *Ph.D Thesis*, SLAC-255, Appendix F.

Gross, E. and Vitells, O. 2010, *Eur. Phys. J. C*, **70**, 525–530.

Hastie, T., Tibshirani, R. and Friedman, J. 2009, *The Elements of Statistical Learning*, Springer.

James, F. 2007, *Statistical Methods in Experimental Physics*, World Scientific.

Kölbig, K. S. and Schorr, B. 1984 *Comp. Phys. Comm.*, **31**, 97, erratum: *ibid* 2008 **178**, 972.

Kendall, M. and Stuart, A. 1979, *The Advanced Theory of Statistics (Vols. 1, 2, and 3)*, Charles Griffin & Company Limited.

Kirkup, L. and Frenkel, B. 2006, *An Introduction to Uncertainty in Measurement*, Cambridge University Press.

Knuth, D. E. 1998, *Art of Computer Programming*, Addison-Wesley.

Landau, L. 1944, *J. Phys. USSR*, **8**, 201; see also W. Allison and J. Cobb 1980, *Ann. Rev. Nucl. Part. Sci.*, **30**, 253.

Lindley, D. V. and Scott, W. F. 1995, *New Cambridge Statistical Tables*, Cambridge University Press.

MacKay, D. 2011, *Information Theory, Inference, and Learning Algorithms*, Cambridge University Press.

Mars Climate Orbiter 1998, see http://mars.jpl.nasa.gov/msp98/orbiter/ for details.

NA62 Collaboration 2010, *Technical Design Report,* NA62-10-07.

Oreglia, M. J. 1980, *Ph.D Thesis* SLAC-236, Appendix D.

Press, W. H. *et al.* 2002, *Numerical Recipes in C++*, Cambridge University Press.

Quinn, H. and Nir, Y. 2007, *The Mystery of the Missing Antimatter*, Princeton University Press.

Rojas, R. 1996, *Neural Networks: A Systematic Introduction*, Springer.

Sahami, M., Dumais, S., Heckerman, D. and Horvitz, E. 1998, *Learning for Text Categorization: Papers from the 1998 Workshop*, AAAI Technical Report WS-98-05.

Silvia, D. S. and Skilling, J. 2008, *Data Analysis, a Bayesian Tutorial*, Oxford University Press.

Skwarnicki, T. 1986, *Ph.D Thesis*, DESY F31-86-02, Appendix E.

Somov, A. *et al.* 2007, *Phys. Rev. D*, **76**, 011104.

Srivastava, A. and Sahami, M. 2009, *Text Mining: Classification, Clustering, and Applications*, Chapman and Hall/CRC.

Wilks, S. S. 1937, *Ann. of Math. Stat.*, **9**, 60–62.

Index